Fatty Acids and Lipids: Biological Aspects

Volume Editors

Claudio Galli
University of Milan

Artemis P. Simopoulos
The Center for Genetics, Nutrition and Health, Washington, D.C.

Elena Tremoli
University of Milan

44 figures and 42 tables, 1994

Basel · Freiburg · Paris · London · New York · New Delhi · Bangkok · Singapore · Tokyo · Sydney

World Review of Nutrition and Dietetics

Library of Congress Cataloging-in-Publication Data
Fatty acids and lipids: biological aspects / volume editors, Claudio Galli,
Artemis P. Simopoulos, Elena Tremoli.
(World review of nutrition and dietetics; vol. 75)
Includes bibliographical references and index.
1. Fatty acids in human nutrition – Congresses. 2. Fatty acids – Metabolism – Congresses.
I. Galli, Claudio. II. Simopoulos, Artemis P., 1933–. III. Tremoli, Elena. IV. Series.
[DNLM: 1. Fatty Acids – metabolism – congresses. 2. Infant Nutrition – physiology – congresses. W1 WO898 v. 75 1994 / QU 90 F253 1994]
ISBN 3-8055-5959-3 (alk. paper)

Bibliographic Indices. This publication is listed in bibliographic services, including Current Contents® and Index Medicus.

Drug Dosage. The authors and the publisher have exerted every effort to ensure that drug selection and dosage set forth in this text are in accord with current recommendations and practice at the time of publication. However, in view of ongoing research, changes in government regulations, and the constant flow of information relating to drug therapy and drug reactions, the reader is urged to check the package insert for each drug for any change in indications and dosage and for added warnings and precautions. This is particularly important when the recommended agent is a new and/or infrequently employed drug.

Printed in Switzerland on acid-free paper by Thür AG Offsetdruck, Pratteln
ISBN 3–8055–5959–3

Fatty Acids and Lipids: Biological Aspects

World Review of Nutrition and Dietetics

Vol. 75

Basel · Freiburg · Paris · London · New York ·
New Delhi · Bangkok · Singapore · Tokyo · Sydney

Contents

Enzymes of PUFA Metabolism and Oxidation

Fatty Acid Oxidation

Fatty Acids and Cell Signalling

Fatty Acids and Human Physiology

Maternal and Infant Nutrition

Mechanisms of Accretion of Polyunsaturates in the Nervous System

PUFA and Natural Antioxidants

Isomeric Fatty Acids

International Society for the Study of Fatty Acids and Lipids

Congress Organization and Sponsors

Honorary President
A. Leaf (USA)

Chairpersons
J. Dyerberg (Denmark)
C. Galli (Italy)
A.P. Simopoulos (USA)

Scientific Secretary
E. Tremoli (Italy)

International Advisory Board
R.G. Ackman (Canada), P. Avogaro (Italy), S.M. Barlow (UK), N.G. Bazan (USA), K.S. Bjerve (Norway), W.E. Connor (USA), M.A. Crawford (UK), G. Crepaldi (Italy), R. De Caterina (Italy), G. De Gaetano (Italy), E. Fedeli (Italy), G.A. Feruglio (Italy), R.A. Gibson (Australia), M. Giovannini (Italy), S.M. Grundy (USA), H. Hansen (Denmark), G. Hornstra (The Netherlands), D. Karyadi (Indonesia), M.B. Katan (The Netherlands), R.R. Kifer (USA), M. Lagarde (France), F. Leighton (Chile), M. Mancini (Italy), R. Martin (USA), P.J. Nestel (Australia), A. Nordoy (Norway), G. Noseda (Switzerland), R. Paoletti (Italy), S. Renaud (France), R.A. Riemersma (UK), N. Salem (USA), A. Sinclair (Australia), C.R. Sirtori (Italy), A.A. Spector (USA), H. Sprecher (USA), Y. Tamura (Japan), P.C. Weber (Germany)

Local Organizing Committee
S. Colli (Italy)
C. Fragiacomo (Switzerland)
F. Marangoni (Italy)
A. Petroni (Italy)

Under the Auspices of
American Association for World Health
American Institute of Nutrition
American Society for Clinical Nutrition
The Center for Genetics, Nutrition and Health
Fondazione Giovanni Lorenzini
Giovanni Lorenzini Medical Foundation
Institute of Pharmacological Sciences, University of Milan
Italian Society of Pharmacology
Nutrition Foundation of Italy
Pan American Health Organization
Swiss League Against Cancer
Ticino Canton League Against Cancer

The Organizing Committee is grateful to all the contributors that have generously supported the organization:
Farmitalia Carlo Erba (Milan, Italy)
F. Hoffmann-La Roche AG (Basel, Switzerland)
Scotia Pharmaceuticals Ltd (Guildford, UK)
Società Prodotti Antibiotici SpA (Milan, Italy)
Swiss League Against Cancer (Switzerland)
Ticino Canton League Against Cancer (Switzerland)

and

Cambridge Isotope Laboratories (Woburn, Mass., USA)
Council for Responsible Nutrition (Washington, D.C., USA)
Glaxo, Inc. (Research Triangle Park, N.C., USA)
Grand Public SA (Paris, France)
Hershey Foods Corporation (Hershey, Pa., USA)
Hoffmann-La Roche Inc. (Nutley, N.J., USA)
Lichtwer Pharma (Berlin, Germany)
Malaysian Palm Oil Promotion Council (Kuala Lumpur, Malaysia)
Mead Johnson Nutritional Group-Bristol-Myers Squibb Company (Evansville, Ind., USA)
Merck & Co., Inc. Research Laboratories (Rahway, N.J., USA)
Merck Sharp & Dohme Research Laboratories (Rahway, N.J., USA)
Monsanto Company (St. Louis, Mo., USA)
National Institute on Alcohol Abuse and Alcoholism (NIAAA) (Bethesda, Md., USA)
National Institute of Child Health and Human Development (NICHD) (Bethesda, Md., USA)
Nestec Ltd Research Center/Nestlé (Lausanne, Switzerland)
Nestlé SA (Vevey, Switzerland)
Novex Pharma Ltd (Henley-on-Thames, Oxon, UK)
Nu-Chek-Prep, Inc. (Elysian, Minn., USA)
Nutricia (Zoetermeer, The Netherlands)
Pronova-Biomed (Lysaker, Norway)
Ross Laboratories (Columbus, Ohio, USA)
Sandoz Pharmaceuticals Corporation (East Hanover, N.J., USA)
Sanofi Winthrop AG (Basel, Switzerland)
The Country Hen (Hubbardston, Mass., USA)
The OmegaSource Corporation (Burnsville, Minn., USA)
Unilever Research Laboratories (Vlaardigen, The Netherlands)
US Department of Agriculture (USDA) (Beltsville, Md., USA)
Wyeth-Ayerst Research (Philadelphia, Pa., USA)

Preface

The Proceedings of the 1st International Congress of the International Society for the Study of Fatty Acids and Lipids (ISSFAL), held June 30–July 3, 1993, in Lugano, Switzerland, are presented in two volumes – volume 75 is entitled 'Fatty Acids and Lipids: Biological Aspects' and volume 76 is entitled 'Effects of Fatty Acids and Lipids in Health and Disease'.

First a brief background about ISSFAL. In March 1990, the 2nd International Conference on the Health Effects of ω3 Polyunsaturated Fatty Acids in Seafoods was held in Washington, D.C. (volume 66 in this series). It was recognized at that time that the field of research on ω3 polyunsaturated fatty acids (PUFA) had considerably advanced to the point that research on the role of *all* the fatty acids and lipids needed to expand. Therefore, it was decided that what was needed was one organization that would bring together scientists and clinical investigators interested in the role of fatty acids and lipids in health and disease. The name of the Society reflects exactly that, the International Society for the Study of Fatty Acids and Lipids – ISSFAL. The Organizing Committee for the Society consisted of Drs. Robert G. Ackman (Canada), Stuart Barlow (UK), Michael Crawford (UK), Jorn Dyerberg (Denmark), Claudio Galli (Italy), Robert Kifer (USA), William E.M. Lands (USA), Alexander Leaf (USA), Federico Leighton (Chile), Roy Martin (USA), Paul Nestel (Australia), Arne Nordøy (Norway), Rodolfo Paoletti (Italy), Serge Renaud (France), Artemis P. Simopoulos (USA), and Peter C. Weber (Germany). Dr. Leaf was appointed President, Dr. Dyerberg, Vice President, and Dr. Simopoulos, Secretary/Treasurer. In March 1991, ISSFAL was established as a nonprofit tax-exempt organization in the Commonwealth of Massachusetts (USA). The stated purpose of the Society is to increase understanding through research and education of the role of dietary fatty acids and lipids in health and disease.

The ISSFAL held its first congress on 'Fatty Acids and Lipids from Cell Biology to Human Disease' in Lugano, Switzerland, June 30–July 3, 1993. The Scientific Secretariat was headed by Drs. Claudio Galli and Elena Tremoli of the University of Milan, and the Organizing Secretariat was headed by Drs. Elena Columbo and Emanuela Folco of the Fondazione Giovanni Lorenzini in Milan, Italy. Four hundred and fifty persons from 35 countries attended, and a total of 332 abstracts consisting of state-of-the-art reviews, critiques and new data were presented at 13 plenary lectures, 10 major symposia, 10 oral communication sessions, and 14 poster sessions, covering topics such as Intracellular Communication; A New Look at Fatty Acids as Signal Transducing Molecules; ω3 PUFA in the Regulation of Cytokine Synthesis; The Role of Fatty Acids during Pregnancy and Lactation; Fatty Acids and Human Physiology; Cardiovascular System; Hypertension; Diabetes; Cancer; Inflammation and Immunology; PUFA and Antioxidants; a round table discussion on The Future of Fatty Acids in Human Nutrition, Health and Policy Implications, and a final session in which summary statements were presented by the session chairmen for general discussion. During this final session it was recommended that trans fatty acids should be labelled as such and should not be included with other fatty acids, such as saturates or PUFA. The need for standardization of fatty acid and lipid nomenclature was recognized, and at the Board Meeting of ISSFAL, a subcommittee was appointed by Dr. Leaf to develop a position paper.

These Proceedings consist of two volumes. Volume 75 includes a summary of the round table discussion on the Future of Fatty Acids in Human Nutrition: Health and Policy Implications; and the Health Message Statement on fatty acids developed by a subcommittee of the ISSFAL Board and approved by a majority of the Board members. Following are the papers presented at the sessions on Enzymes of PUFA Metabolism and Oxidation; Fatty Acid Oxidation; Fatty Acids and Cell Signalling; Fatty Acids and Human Physiology; Maternal and Infant Nutrition; Mechanisms of Accretion of Polyunsaturates in the Nervous System; PUFA and Natural Antioxidants, and Isomeric Fatty Acids.

Volume 76 includes the papers presented at the sessions on Fatty Acids and Cardiovascular System; ω3 and ω6 Fatty Acids, Lipids and Lipoproteins; ω3 Fatty Acids and Thrombosis; Fatty Acids and Cancer; Inflammation and Immunology; Essential Fatty Acids; Pregnancy and Pregnancy Complications, and Clinical Trials with ω3 Fatty Acids.

These Proceedings and the congress abstracts contain the most up-to-date information on the physiological and metabolic aspects of fatty acids and lipids in health and disease, in the form of reviews, new data, and state-of-the-art papers on the role of ω3 fatty acids and their relationship to ω6, ω9, saturated fats and trans fatty acids in biomedical research and in clinical investigations. The Proceedings should be of interest to biomedical researchers in academia, industry and government, including clinical investigators, physiologists, biochemists, nutritionists, dietitians, and policy-makers.

Artemis P. Simopoulos, MD
Claudio Galli, MD
Elena Tremoli, PhD

....

Round Table

The Future of Fatty Acids in Human Nutrition: Health and Policy Implications

Artemis P. Simopoulos

This session was originally proposed by Dr. Michael Crawford and was to be cochaired by Dr. Crawford and Dr. Simopoulos. Unfortunately, Dr. Crawford was unable to participate because of personal reasons. In his absence, Dr. Simopoulos read Dr. Crawford's statement:

'The International Society for the Study of Fatty Acids and Lipids (ISSFAL) must take a leadership role in raising the political/scientific awareness on fatty acids for the following reasons:

(1) The failure of the XV International Congress of Nutrition (IUNS), to be held in Adelaide, Australia, September 26 to October 1, 1993, to take on board fatty acid issues;

(2) The failure of the Food and Agricultural Organization of the United Nations (FAO) to include issues on fatty acids in their agenda. FAO and WHO held an Expert Consultation Meeting on Fats and Oils in Human Nutrition, in Rome, October 13–26, 1993, specifically devoted to fatty acids;

(3) Evidence that fatty acids are not being considered seriously in the political arenas where they matter most and often far more than the issues they repeatedly debate; and

(4) Priorities for research funding are outdated by two decades of new knowledge on fatty acids.'

Participants in this session included Drs. Kristian Bjerve, Peter Bourne, Raffaele De Caterina, Jacqueline Dupont, Bertold Koletzko and Alexander Leaf. The following topics were discussed both from health and policy aspects.

Trans Fatty Acids

Data on this subject were presented at the sessions on Isomeric Fatty Acids; Maternal and Infant Nutrition, and on Essential Fatty Acids, Pregnancy, and Pregnancy Complications. The safety of trans fatty acids during pregnancy and the perinatal period is in question. As a first step in this direction, a regulation has been proposed for the European Community to restrict the maximum content of trans fatty acids in infant formulas to no more than 4% of total fat.

Unlike the results of animal studies in which trans fatty acids appear not to cross the placenta, in human studies trans fatty acids cross the placenta and interfere with the growth of the fetus. The higher the cord blood concentration, the lower the fetal weight. Trans fatty acids interfere with the desaturation and elongation of essential fatty acids. Trans fatty acids behave much like saturated fats in terms of their effects on LDL cholesterol. They raise LDL, while decrease HDL. They also increase triglycerides and Lp(a), a most atherogenic and thrombogenic lipoprotein. In Western diets, trans fatty acids are estimated to contribute 4–15% energy. It is essential that governments develop accurate information on the composition of fatty acids and lipids in the food supply. Most trans fatty acids result from the hydrogenation of vegetable oils (and to a lesser extent of fish oils). It is a challenge for the manufacturers of processed foods to reduce the content of trans fatty acids and to develop alternatives. Eventually, diets should not contain trans fatty acids as a result of hydrogenation.

Fatty Acids in Perinatal Nutrition

It is now unanimously agreed that docosahexaenoic acid (DHA) is essential for the normal development of the premature. Two studies have already confirmed that DHA is also essential for the full-term newborn. Infants of low birth weight fed formula devoid of DHA have functional deficits. Arachidonic acid (AA) present in human milk has been associated to quality of growth and development. For that reason it is essential that dietary intervention does not interfere with AA availability and does not lead to its depletion. It is necessary to supplement artificial foods and formula for preterm infants with DHA, as was recommended in 1991 by the Nutrition Committee of the European Society for Pediatric Gastroenterology and Nutrition (ESPGAN).

Maternal disease and diet appear to have significant effects on milk composition; there is also considerable individual variation and the effects of the various long chain polyunsaturated fatty acids (LCPUFA) need to be precisely established. Dietary trans fatty acid intake interferes with desaturation and elongation of essential fatty acids (EFA), which indicates that in the absence of trans fatty acids, the quantity of AA and DHA in human milk might be higher. Therefore, the amounts of DHA and AA in the breast milk should be considered as the minimal amounts to be incorporated into the preterm infant's formula. For the full-term human, milk is the ideal food and every effort should be made that mothers breast-feed and it is made easy for them to breast-feed. Full-term newborns should be given formula only under extreme circumstances. It is important that research is carried out on the EFA requirements during both pregnancy and lactation since current Western diet and infant formula are not necessarily optimal, even for the full term. In the meantime, the composition of human milk could serve as a standard for all infant formula.

Dietary Requirements for Adults

Studies on persons deficient in long chain omega-3 fatty acids require a daily intake of 300–400 mg of DHA and EPA (about 0.4% of energy) or 900–1,100 mg of alpha-linolenic acid to achieve a status similar to the median levels in the Norwegian population. It is not known if the median level is also the optimal level. Either fish or fish oil could be recommended depending on taste and individual preference. Although some studies show that the effects of DHA and eicosapentaenoic acid (EPA) are not equivalent in certain disease states, for healthy people most likely their effects are equivalent. In order to precisely define the requirements of the fatty acids, there is a need for better standardization of the methods and of the reporting of clinical investigations. The use of mass units such as g/day or mol/day would make comparisons easier. There is also a need to better design experiments building on the techniques of molecular biology. Food composition tables are needed that provide accurate information on the lipid and fatty acid content of food. Because few studies present complete analyses of the diets fed to subjects, this lack of dietary information contributes to difficulties in interpreting the results.

Fatty Acids in Cardiovascular Disease

Most of the epidemiologic studies, animal experiments and clinical investigations have been carried out on this topic. The new information on the evolution and development of atherosclerotic process indicates a series of events that take place. Initiation of the lesion is brought about by endothelial activation (damage), followed by vascular adhesion molecules which recruit monocytes to the intima, followed by progression of the lesion, fissuring, platelet activation, thrombosis and ischemia, electrical instability and arrhythmia or ventricular fibrillation (the main cause of sudden death). Omega-3 fatty acids have been shown to interfere in all the steps of the process or chain of events, whereas no other natural or pharmacological substance has been shown to have so many different effects. However, when one considers the evolutionary aspects of diet and the changes that have taken place in the Western diet, it is tempting to conclude that in terms of abnormalities in lipid metabolism, atherosclerosis is not only a disease of maladaptation between the genetic profile and increases in saturated fat and cholesterol, but a deficiency disease of omega-3 fatty acids as well.

Fatty Acids and Cancer

Most of the research has been on animal studies. In such studies linoleic acid promotes tumor growth, whereas alpha-linolenic acid decreases the size and number of tumors. In cell cultures, omega-3 fatty acids decrease the expression of oncogenes. Epidemiologic studies show an inverse relationship between alpha-linolenic acid intake and death from all cancers. Similarly, fish-eating populations have less cancer than nonfish-eating populations. This is an area where a lot more research is needed involving clinical trials, and cell culture studies, in order to improve our understanding of the cellular mechanisms involved in cell proliferation, oncogene function, and cancer development.

Incorporating Scientific Knowledge in Policy Development and Public Education: The Role of Governments and International Organizations

Government policy should be based on the most sound and current scientific knowledge. This issue is complex because, in the developed countries, up to now political and commercial interests have been the primary factors in shaping national and international policies instead of nutrition policy being the primary factor to drive agricultural policy.

The improvement of nutrition as a global public health strategy is essential as is the communication between scientists and the public to reach this goal. Governments need to establish commissions on nutrition and food at the highest level of governments (i.e., a Presidential Commission in the US) in order that nutrition policy drives agricultural policy. Strengthening and expansion of the Nutrition and Food Policy Division at the World Health Organization (WHO) and at the Food and Agriculture Organization of the United

Nations (FAO) are essential steps in order for the international community to take action.

The International Society for the Study of Fatty Acids and Lipids (ISSFAL) is the proper organization to develop a series of clear policy recommendations that could be implemented at a policy level with governmental and intergovernmental organizations. In this way, scientific knowledge could be rapidly transferred into public health benefits for people all over the world.

The ISSFAL Board responded to the challenge and as a first action developed a health message statement on omega-3 fatty acids and cardiovascular disease that appears on the following page.

Health Message Statement

Prevention and treatment of coronary heart disease (CHD) in the community require changes in lifestyle risk factors which are reversible. These include smoking and physical inactivity. Furthermore, faulty dietary habits can lead to high blood lipid levels, contribute to high blood pressure and increase the tendency for thrombosis. Extensive scientific data support that several of these diet-related risk factors can be substantially diminished by the omega-3 fatty acids found in fatty fish. Dietary intake of such omega-3 fatty acids higher than is usual in Western diets is advisable in order to reduce the risk of CHD.

This health message statement was developed by a subcommittee of the Board of Directors of the International Society for the Study of Fatty Acids and Lipids (ISSFAL) and approved by the majority of the Directors in August 1993.

Galli C, Simopoulos AP, Tremoli E (eds): Fatty Acids and Lipids: Biological Aspects.
World Rev Nutr Diet. Basel, Karger, 1994, vol 75, pp 1–7

Intercellular Communication in Fatty Acid Metabolism[1]

Howard Sprecher, Devanand Luthria, Michelle Geiger, B. Selma Mohammed, Meri Reinhart

Department of Medical Biochemistry, The Ohio State University, Columbus, Ohio, USA

Unless the type of fat in the diet is changed, the fatty acid composition of membrane lipids is relatively independent of the percent of calories provided by fat. The incorporation of fatty acids into phospholipids must thus be regulated via some type of coordinated mechanism with the content of polyunsaturated fatty acids (PUFA) from dietary precursors. It is generally accepted that desaturation of 18:2ω6 is the rate-limiting step for the synthesis of arachidonate. Changes in arachidonate levels of lipids in disease states are sometimes related to and assumed to be correlated with altered Δ6-desaturase activity [1]. However, it is possible that the compositional differences are due to alterations in the regulation of phospholipid rather than PUFA biosynthesis. In theory, the amount of arachidonate synthesized may be modified by adding acids beyond the Δ6-desaturase step to the diet. The role that liver plays in providing PUFA to extrahepatic tissues remains a topic of interest. By feeding a series of deuterated fatty acids to rats we have been able to quantify and compare the specific activity of 18:2ω6 and 20:4ω6 in liver, heart and kidney to determine what regulates the synthesis of arachidonate in liver and the possible transport of arachidonate to extrahepatic tissues.

In 1991 we reported that liver did not have the capacity to metabolize 22:5ω3 to 22:6ω3 via a putative Δ4-desaturase [2]. We have recently shown that 22:5ω6 is synthesized via an analogous pathway that also is independent of Δ4-desaturase. This finding however requires that 18:2ω6, 18:3ω3, 24:4ω6 and 24:5ω3 are all substrates for desaturation at position 6. In the second part of

[1] These studies were supported by NIH Grant DK 20387.

Table 1. The molar percent of deuterated linoleate and arachidonate in liver, heart and kidney phospholipids after feeding 17,17,18,18-d_4-ethyl linoleate

	Linoleate	Arachidonate
Liver	57	34
Heart	41	9
Kidney	45	13

this brief review we will describe some of the recent studies we have done on the factors regulating 22:5ω6 and 22:6ω3 biosynthesis with a focus on the efforts we have made to determine whether rat liver might contain chain-length specific Δ6-desaturases.

Metabolism of Deuterated Fatty Acids

These studies were all carried out by maintaining rats on a modified AIN-76 diet in which the dietary lipid was a mixture of pure ethyl esters. These diets were fed for 4 weeks after which time a single ethyl ester was replaced with an equal amount of the deuterated ethyl ester. After 4 days, the animals were sacrificed, the phospholipids were separated from neutral lipids and the total phospholipids were saponified. The free fatty acids were derivatized by reaction with N-methyl-N-(t-butyldimethylsilyl)trifluoroacetamide containing 1% t-butyldimethylchlorosilane. The derivatized acids were analyzed by gas chromatography mass spectrometry and the appropriate M-57 ions of endogenous and deuterated linoleate and arachidonate were integrated.

The percent of fat in the diet of all groups of animals was 3.3% and in group 1 it consisted of 2.1% oleate, 1.0% linoleate and 0.2% linolenate by weight of the total diet. After 4 weeks on this diet, the linoleate was replaced by an equal amount of 17,17,18,18-d_4-linoleate. The results in table 1 compare the molar amounts of deuterated linoleate and arachidonate in liver, heart and kidney phospholipids. The observed different specific activities between esterified linoleate and arachidonate in liver could possibly be explained in two ways. The lower specific activity of arachidonate would be observed if the arachidonate-containing phospholipids had a slower turnover rate than did those containing 18:2ω6. This is unlikely since Schmid et al. [3] have recently shown, with stable isotope methodology, that the turnover rates of linoleate- and arachidonate-containing molecular species in ethanolamine- and choline-containing phosphoglycerides were similar. It seems more likely that dietary linoleate mixes with endogenous linoleate to establish a fatty acid pool the specific activity of which may be regulated by the amount of linoleate in the diet. The specific activity of linoleate in phospholipids is thus a reflection of the specific activity of linoleate in the free fatty acid pool which may be directly related to dietary linoleate intake. The lower specific activity of arachidonate suggests that the ongoing synthesis of arachidonate is, by itself, insufficient to meet the needs for the amount of arachidonate required for membrane lipid biosynthesis. Steady-state levels of esterified arachidonate can then only be maintained by considerable recycling of esterified arachidonate. This hypothesis implies that desaturation of linoleate by the Δ6-desaturase is not only rate limiting for arachidonate biosynthesis, but also regulates the amount of newly formed arachidonate available for membrane lipid synthesis.

The data in table 1 show that the specific activity of esterified linoleate in heart and kidney phospholipids is similar but lower than in liver. It would appear that there are adequate d_4-linoleate levels in these tissues for metabolism to arachidonate, yet the specific activity of arachidonate is much lower than in liver. We were unable to observe any synthesis of arachidonate when heart cardiomyocytes were incubated with radioactive 18:2ω6, 18:3ω6 or 20:3ω6 [4]. There remains some debate as to whether kidney microsomes have the ability to desaturate fatty acids at positions 5 or 6 [5]. Our labeling pattern suggests that linoleate is probably metabolized to arachidonate in liver which is then transported to these two tissues. These results would support the findings of Lefkowith et al. [6] showing that arachidonate in liver was exported to both heart and kidney.

To determine whether products of the Δ6-desaturase either down-regulate the amount of linoleate metabolized to arachidonate or modify the amount of arachidonate synthesized, another type of feeding experiment was designed. In group 2, rats were again fed a diet containing 3.3% fat and it consisted of 1.9% 18:1ω9, 1% 18:2ω6, 0.2% 18:3ω3 and 0.2% 18:3ω6. After 4 days, the rats in group 2A were fed d_4-18:2ω6 to determine whether 18:3ω6 modified the amount of linoleate metabolized to 20:4ω6. In group 2B,17,17,18,18-d_4-18:3ω6 was fed to quantify how much of this acid was metabolized to 20:4ω6. By adding the molar amounts of deuterated (d_4) 20:4ω6 made in groups 2A and 2B, we were able to quantify the fractional molar amount of d_4-20:4ω6 made and esterified into phospholipids by the combined metabolism of 18:2ω6 and 18:3ω6. An identical protocol was followed in groups 3A and 3B and 4A and 4B, except the variables were 20:3ω6 and 20:4ω6. These acids contained four deuterium atoms – i.e. 19,19,20,20-d_4. The addition of 18:3ω6, 20:3ω6 and 20:4ω6 to the diets did not affect the molar amount of d_4-18:2ω6 esterified in phospholipids (data not shown). The other conclusions of these studies, as they

Table 2. The molar percent of deuterated arachidonic acid in liver phospholipids as influenced by feeding ω6 acids to rats

Group 1	Group 2A	Group 2B
d_4-18:2ω6	d_4-18:2ω6 d_0-18:3ω6	d_0-18:2ω6 d_4-18:3ω6
34	27	25

Group 1	Group 3A	Group 3B
d_4-18:2ω6	d_4-18:2ω6 d_0-20:3ω6	d_0-18:2ω6 d_4-20:3ω6
34	24	22

Group 1	Group 4A	Group 4B
d_4-18:2ω6	d_4-18:2ω6 d_0-20:4ω6	d_0-18:2ω6 d_4-20:4ω6
34	24	26

relate to liver phospholipids, are summarized in table 2. The addition of 18:3ω6, 20:3ω6 and 20:4ω6 always depressed the molar amount of d_4-20:4ω6 esterified in phospholipids – i.e. compare groups 2A, 3A and 4A with group 1. The sum of the molar amount of d_4-20:4ω6 in phospholipids in groups 2A and 2B, 3A and 3B and 4A and 4B was always considerably greater than group 1. Dietary supplements of 18:3ω6, 20:3ω6 and 20:4ω6 do thus increase the amount of 20:4ω6 that was made. Collectively, these data show that dietary supplements of 18:3ω6, 20:3ω6 or 20:4ω6 depress the metabolism of 18:2ω6 to 20:4ω6. However, the total amount of arachidonate esterified into phospholipids as made from 18:2ω6, plus that made when 18:3ω6, 20:3ω6 or 20:4ω6 were added to the diet always exceeded that made when 18:2ω6 was the only ω6 dietary acid. These data suggest that if there is a necessity to increase 20:4ω6 for membrane lipid biogenesis, it can be accomplished by feeding supplements of ω6 acids beyond the Δ6-desaturase step. However, it is extremely important to stress that the effects of adding 18:3ω6, 20:3ω6 or 20:4ω6 to the diet are not easy to quantify. In these dietary studies the actual amount of 20:4ω6 esterified in liver phospholipids was the same when only

Table 3. Rate of esterification when acyl-CoA derivatives were incubated with rat liver microsomes and 1-acyl-sn-glycero-3-phosphocholine

Substrate	Rate, nmol/min/mg microsomal protein
20:4ω6 CoA	53
22:4ω6 CoA	6
22:5ω6 CoA	50
24:4ω6 CoA	0
24:5ω6 CoA	0

linoleate was fed versus when other ω6 acids were added to the diet. In this regard, compositional analysis is not a totally accurate measure of PUFA metabolism in vivo. Finally, it should be noted that the ratio of 18:2ω6 to the other ω6 acids was always 5:1. The molar fraction of d_4-20:4ω6 made and esterified into liver phospholipids was the same when 18:3ω6 or 20:3ω6 were fed as when 20:4ω6 itself was provided – i.e. compare groups 2B, 3B and 4B. With all three ω6 acids their contribution to esterified 20:4ω6 was about the same as for linoleate – i.e. compare groups 2A and 2B, 3A and 3B, and 4A and 4B – even though the diet contained 5 times as much 18:2ω6.

Synthesis of 22:5ω6 and 22:6ω3

The types of experiments carried out to show that 22:5ω6 was made via a pathway independent of a Δ4-desaturase were similar to those we used to determine how 22:5ω3 was metabolized to 22:6ω3 [2]. When 22:4ω6 was incubated under desaturating conditions with microsomes, it was not metabolized to 22:5ω6. When malonyl-CoA was included in incubations, it was possible to detect radioactive 24:4ω6 and 24:5ω6. With hepatocytes it was possible to show that [1-^{14}C]22:4ω6, [3-^{14}C]24:4ω6 and [3-^{14}C]24:5ω6 were all metabolized to 22:5ω6.

A puzzling observation with the hepatocyte studies was the finding that when 24-carbon ω3 and ω6 acids were the substrates, only small amounts were esterified directly into phospholipids. The results in table 3 compare the rates of esterification when appropriate acyl-CoAs were incubated with rat liver microsomes and 1-acyl-sn-glycero-3-phosphocholine using a continuous spectral assay [7]. With this methodology it was not possible to detect acylation of either 24-carbon ω6 acid. We now hypothesize that 22:4ω6 is chain elongated

Table 4. Competitive interactions for desaturation at position 6 between 18:3ω3 and 24:5ω3 using rat liver microsomes

Substrates		Product	Substrates		Product
nmol [3-^{14}C]24:5ω3	nmol 18:3ω3	nmol 24:6ω3 produced	nmol [1-^{14}C]18:3ω3	nmol 24:5ω3	nmol 18:4ω3 produced
80	–	7.7	80	–	9.2
80	40	4.4	80	40	7.9
80	80	2.6	80	80	7.4
80	120	1.6	80	120	6.8

Table 5. Effect of dietary ω3 acids on the desaturation of 18- and 24-carbon acids at position 6 by rat liver microsomes

Diet	Substrate, nmol of product formed			
	[1-^{14}C]18:2ω6	[1-^{14}C]18:3ω3	[3-^{14}C]24:4ω6	[3-^{14}C]24:5ω3
A	5.8	5.9	4.8	5.5
B	7.0	8.4	4.1	6.1
C	1.9	3.0	2.7	2.9
D	4.2	6.4	4.5	5.4

All diets contained 5% fat. The fat in diet A contained 33% linoleate and 55% oleate. In diets B, C and D, 10% of 18.3ω3, 20:5ω3 and 22:6ω3 replaced an equal amount of oleate.

to 24:4ω6 which is then desaturated in the microsome to 24:5ω6. This acid may then move to peroxisomes where it is chain shortened to 22:5ω6 after which it moves back to the endoplasmic reticulum for esterification.

Two types of experiments were carried out to determine if rat liver microsomes might contain chain-length specific acyl-CoA Δ6-desaturases. In the first experiment, an enzyme-saturating level of.[3-^{14}C]24:5ω3 was incubated together with increasing amounts of 18:3ω3 to determine whether it depressed 24:5ω3 biosynthesis. In the reciprocal experiment an enzyme-saturating level of [1-^{14}C]18:3ω3 was coincubated with increasing amounts of 25:5ω3. The results, as tabulated in table 4, show that 18:3ω3 was a potent inhibitor of 24:5ω3 desaturation. Conversely, 24:5ω3 was a rather poor inhibitor of 18:3ω3 desaturation. If these microsomal studies can be extrapolated to whole-body metabolism, they suggest that 22:6ω3 biosynthesis might be depressed by high dietary levels of 18:3ω3. In the second experiment, we fed rats a diet in which 5% of the diet was fat and 33% of the fatty acids in the fat were linoleate. Rats were then fed the same diet but 10% of the oleate in the diet was replaced by pure 18:3ω3, 20:5ω3 or 22:6ω3. The rationale for these studies was to determine whether Δ6-desaturase activity was down-regulated when it was not required to produce 22:6ω3. The results in table 5 show that only when 20:5ω3 was added to the diet was there a marked depression in Δ6-desaturase activity. However, this depressed activity was observed for all four substrates. Thus, neither of these studies provided conclusive evidence for the presence of more than one Δ6-desaturase. Clearly, there is glaring need to isolate the proteins that carry out both the desaturation and chain elongation of PUFA to determine how many of them are involved in this process and what regulates their synthesis. The need for this type of study is even more apparent in light of recent studies showing that there is more than one form of the Δ9-desaturase [8].

References

1 Holman RT: A long scaly tale – The study of essential fatty acids deficiency at the University of Minnesota; in Sinclair A, Gibson R (eds): Essential Fatty Acids and Eicosanoids, Champaign, American Oil Chemists Society, 1992, pp 3–17.
2 Voss A, Reinhart MS, Sankarappa S, et al: The metabolism of 7,10,13,16,19-docosapentaenoic acid to 4,7,10,13,16,19-docosahexaenoic acid is independent of a 4-desaturase. J Biol Chem 1991;266:19995–20000.
3 Schmid PC, Johnson SB, Schmid HO: Remodeling of rat hepatocyte phospholipids by selective acyl tumover. J Biol Chem 1991;266:13690–13697.
4 Hagve T-A, Sprecher A: Metabolism of long-chain polyunsaturated fatty acids in isolated cardiac myocytes. Biochim Biophys Acta 1989;1001:338–344.
5 Suneja SK, Nagai MN, Cook L, et al: Do kidney cortex microsomes possess the enzymatic machinery to desaturate and chain elongate fatty acyl-CoA derivatives? Lipids 1991;26:359–363.
6 Lefkowith JB, Flippo V, Sprecher H, et al: Paradoxical conservation of cardiac and renal arachidonate content in essential fatty acid deficiency. J Biol Chem 1985;260:15736–15744.
7 Lands WEM, Inoue M, Sugiura Y, et al: Selective incorporation of polyunsaturated fatty acids into phosphatidylcholine by rat liver microsomes. J Biol Chem 1982;257:14968–14972.
8 Ntambi JM: Dietary regulation of stearoyl-CoA desaturase 1 gene expression in mouse liver. J Biol Chem 1992;267:10925–10930.

Dr. Howard W. Sprecher, Department of Medical Biochemistry, The Ohio State University, 337 Hamilton Hall, 1645 Neil Avenue, Columbus, OH 43210 (USA)

Galli C, Simopoulos AP, Tremoli E (eds): Fatty Acids and Lipids: Biological Aspects.
World Rev Nutr Diet. Basel, Karger, 1994, vol 75, pp 8–15

β-Oxidation of Hydroxyeicosatetraenoic Acids: A Peroxisomal Process[1]

Arthur A. Spector[a, b], *Joel A. Gordon*[b], *Steven A. Moore*[c]

Departments of [a]Biochemistry, [b]Internal Medicine and [c]Pathology,
College of Medicine, University of Iowa, Iowa City, Iowa, USA

Hydroxyeicosatetraenoic acids (HETEs) are arachidonic acid (AA) metabolites containing a single hydroxyl group that are formed primarily by lipoxygenase pathways [1]. The three main forms of HETE that are synthesized by human and animal cells are 5-, 12- and 15-HETE which have hydroxyl groups on carbons 5, 12, and 15, respectively, counting from the carboxyl group as carbon 1. The type and amount of HETE produced by a cell depends on the lipoxygenases that it contains. For example, neutrophils have an active 5-lipoxygenase pathway and are a major source of 5-HETE, platelets produce large quantities of 12-HETE when they are activated, and tracheal epithelial cells and eosinophils are major producers of 15-HETE.

Production and Release from Cells

Figure 1 summarizes our present understanding about the synthesis and trafficking of HETEs in mammalian cells. The AA utilized for HETE synthesis usually is derived from intracellular phospholipids as a result of the activation of phospholipase A_2. Alternatively, the AA probably can be derived from extracellular sources, for some tissues produce HETEs when they are incubated in media containing AA [2]. Lipoxygenases convert the AA to a hydroperoxy-

[1] The studies from the authors' laboratories described in this review were supported by NIH grants HL49264, NS24261, NS01096 and Grant-in-Aid IA-92-GS-37 from the Iowa Affiliate of the American Heart Association.

eicosatetraenoic acid (HPETE). The HPETE can be utilized for the synthesis of eicosanoid mediators such as leukotrienes or lipoxins, may regulate certain intracellular processes [1], or may possibly be secreted and mediate extracellular oxidations. Through the action of glutathione peroxidase or other reductases, the hydroperoxy group of HPETE is converted to a hydroxyl group, forming the corresponding HETE. Most or all of the HETE that is produced is released into the surrounding medium.

Function of HETEs

Although the HETEs have proinflammatory actions, they occur only at relatively high concentrations and are weaker than those produced by the leukotrienes. This led to the view that HETEs are inactivation products of the lipoxygenase pathways and are not important bioactive eicosanoids. As shown in figure 1, however, Stenson and Parker demonstrated that some of the released HETE is reincorporated and esterified into phospholipids. This finding suggested that HETEs may be a new type of autacrine mediator that acts by altering membrane structure and function [3]. Subsequently, HETE transcytosis was demonstrated; the HETE released by one cell was either incorporated or converted to another metabolite by a second type of cell. For example, 12-HETE released by platelets was incorporated by endothelial cells [4] or converted to 12,20-diHETE by neutrophils [5]. This suggested that HETEs also may have paracrine actions.

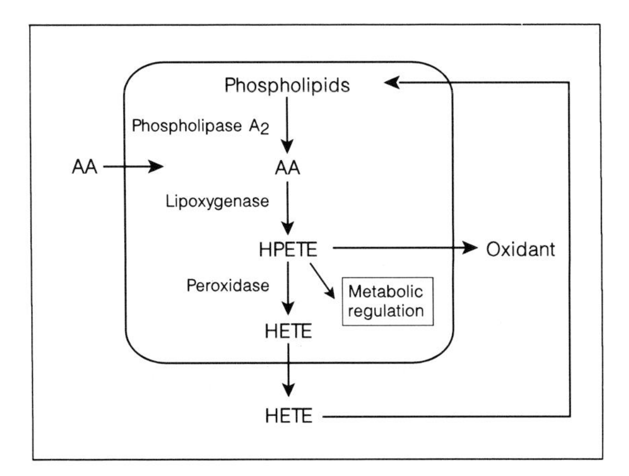

Fig. 1. Synthesis and release of HETEs from mammalian cells.

Subsequent studies demonstrated a number of specific actions of HETEs in different tissues. For example, 12-HETE inhibits smooth muscle proliferation [6] and may be involved in neuronal transmembrane signaling [7], 15-HETE enters the inositol phospholipid (PI) signaling pathway [8–10], and all of the HETEs reduce prostaglandin formation [11–14]. 12-HETE production increases when mouse peritoneal macrophages are converted to foam cells [15], and there is increased 15-lipoxygenase expression in atheroma [16]. The latter findings suggest that 12- and 15-HETE may be involved in the atherogenic process.

HETE Incorporation into Cells

To more fully understand the biological actions of HETEs, we have investigated their interaction with arterial endothelial cells [8, 17], brain microvessel endothelial cells [12], renal tubular epithelial cells [13, 18] and human skin fibroblasts [19]. This work, together with that of other groups [1, 3, 9, 10], provides a general understanding of HETE metabolism. The amounts taken up are 5- > 12- > 15-HETE, and the total incorporation is somewhat less than that of AA. Endothelial and renal epithelial cells grown on micropore filters can take up HETEs from both the apical and basolateral surfaces [8, 13, 18]. HETEs are incorporated into the cell phospholipids. Most of the 5- and 12-HETE uptake is present in choline phosphoglycerides (PC) [1, 12, 17–19], whereas a large amount of the 15-HETE uptake is incorporated into PI [1, 8–10]. Much of the uptake is contained in the membrane fractions of endothelial homogenates, with the microsomes taking up more than the plasma membrane [17]. 15-HETE also is incorporated into endothelial triglycerides [8], and some of the 12-HETE taken up by macrophage foam cells is incorporated into cholesterol esters [20].

Formation of HETE Metabolites

During incubation of [^{3}H]-labeled HETEs with cells, there is a continuing uptake during the first 2–4 h. This is followed by a gradual reduction in the amount present in the cells and a reaccumulation of lipid-soluble radioactivity in the medium [17]. Likewise, following pulse labeling with [^{3}H]HETEs, a substantial release of lipid-soluble radioactivity occurs during the first hour even though no agonist is added to the second incubation medium [8, 12, 17]. Much of the radioactivity released from the cells is derived from phospholipids [8, 17]. These results indicate that much of the newly incorporated HETE is labile and suggest that the presence of an oxygenated fatty acyl chain may perturb the structure of the lipid bilayer and make the phospholipids more susceptible to hydrolysis by phospholipases [1, 3].

Identification of Products

Assay by reverse-phase high-performance liquid chromatography (HPLC) indicated that during incubations with 12- or 15-HETE, several radioactive metabolites with shorter retention times than the HETE gradually accumulated in the extracellular fluid [8, 12–14, 19]. One product predominated, especially during the first 1–2 h. Analysis of the main 12-HETE product by gas-liquid chromatography combined with mass spectrometry indicated that it is 8-hydroxy-4,6*t*,10-hexadecatrienoic acid (16:3[8-OH]) [12, 21]. Similarly, the main metabolite formed from 15-HETE was found to be 11-hydroxy-4,7,9*t*-hexadecatrienoic acid (16:3[11-OH]) [8].

These findings suggested that the metabolites were formed by partial β-oxidation [1, 8, 21]. If 12- and 15-HETE undergo two successive β-oxidations, removing four carbons from the carboxyl end, the 16-carbon products will be formed (fig. 2). In our initial studies, the putative 18-carbon intermediate formed after the first β-oxidation cycle was not detected in appreciable amounts. Subsequent work by others with macrophages and MOLT-4 lymphocytes demonstrated the presence of the 18:3 intermediates, as well as additional metabolites in the β-oxidation pathway [22, 23]. These include 14- and

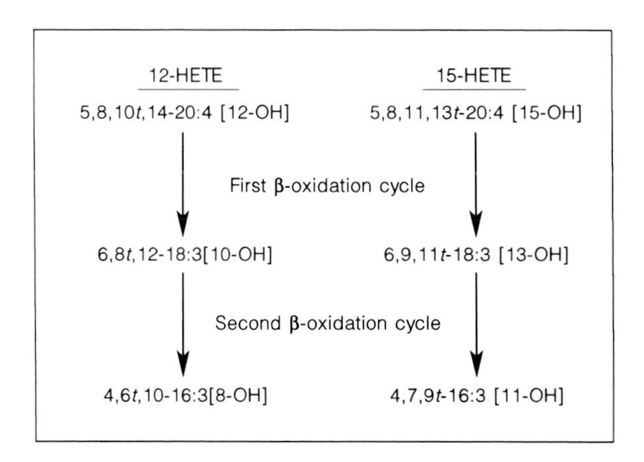

Fig. 2. Pathway for the conversion of 12- and 15-HETE to the 16-carbon monohydroxy metabolites. The two β-oxidation cycles take place in the peroxisomes. These reactions occur with the compounds in the form of CoA derivatives, but the CoA has been omitted from the structures for clarity.

12-carbon products, indicating that some further oxidation beyond the 16-carbon stage occurs in the intact cell. This is consistent with results demonstrating that in the human intestinal CaCo-2 cell line, most of the 15-HETE uptake is converted to short-chain oxidation products, and very little 16:3[11-OH] accumulates [24].

A similar series of metabolites are not formed from 5-HETE. However, a single radioactive product was detected by HPLC when [^{3}H]5-HETE was incubated with porcine coronary artery endothelial cells. This product has the HPLC retention time expected for an 18-carbon monohydroxy intermediate, but it has not as yet been identified [14].

Role of Peroxisomes

Because the formation of the hexadecatrienoic acid metabolites was not inhibited when cells were treated with a mitochondrial β-oxidation inhibitor [19], the possibility of peroxisomal oxidation was investigated. These studies indicated that 12- and 15-HETE were converted to 16:3[8-OH] and 16:3[11-OH], respectively, by normal human skin fibroblasts but not by fibroblasts derived from patients with Zellweger's disease, a peroxisomal deficiency [19]. Similarly, the metabolites were produced by wild-type Chinese hamster ovary (CHO) cells but not by two peroxisome-deficient mutant CHO cell lines, ZR78 and ZR82 [25]. The conclusion that the process is peroxisomal is supported by the results of others demonstrating that 12-HETE is converted to 16:3[8-OH] by isolated liver and kidney peroxisomes [26].

These data indicate that 12- and 15-HETE are degraded through a β-oxidation process in which at least the first two cycles occur in the peroxisomes. In many cells including fibroblasts [19], CHO cells [25], endothelium [8, 12, 14], epithelium [13], macrophages [22] and lymphoma cells [23], the process apparently slows down sufficiently after the first two β-oxidation cycles to allow the 16-carbon monohydroxy intermediate to accumulate and be released into the surrounding fluid. The question of why this intermediate accumulates has not been resolved.

One possibility relates to the fact that 16:3[8-OH] and 16:3[11-OH] contain 4-*cis* unsaturated bonds. A special enzyme, 2,4-dienoyl coenzyme A (CoA) reductase, is required to continue the β-oxidation of unsaturated fatty acids containing 4-*cis* double bonds [27]. This may be a slow reaction relative to the preceding steps in the peroxisomal β-oxidation pathway in many cells, causing the 16:3 intermediate to accumulate to a level where some of it interacts with a thioesterase and is released as the free acid into the extracellular fluid.

Function of Hexadecatrienoic Acids

A critical question is whether 16:3[8-OH] or 16:3[11-OH] accumulate simply because they cannot be oxidized rapidly enough, or whether they are formed because they have biological actions. No specific function for either of the compounds has so far been observed. However, brain microvessel endothelial cells slowly reaccumulate 16:3[8-OH] after it is released [12], and a small amount of 16:3[11-OH] is incorporated into endothelial lipids [8]. Unlike 15-HETE which is incorporated into PI, 16:3[11-OH] is channeled into PC [8]. The fact that the 16:3 metabolites can be taken by cells suggests that they may have a functional effect.

Conclusions

Figure 3 summarizes our current view of HETE metabolism and the role of the β-oxidation process. HETEs can be taken up by many cells. Following formation of the CoA thioester, some of the HETE is incorporated into cell lipids including PC and PI, phospholipids that are involved in membrane signal transduction. The 12- and 15-HETE CoA also can undergo peroxisomal β-oxidation, which may be a mechanism for facilitating the removal of these HETEs from the cell. After two cycles of β-oxidation, a rate-limiting reaction is encountered causing the 16:3 intermediate to accumulate. We suggest that the

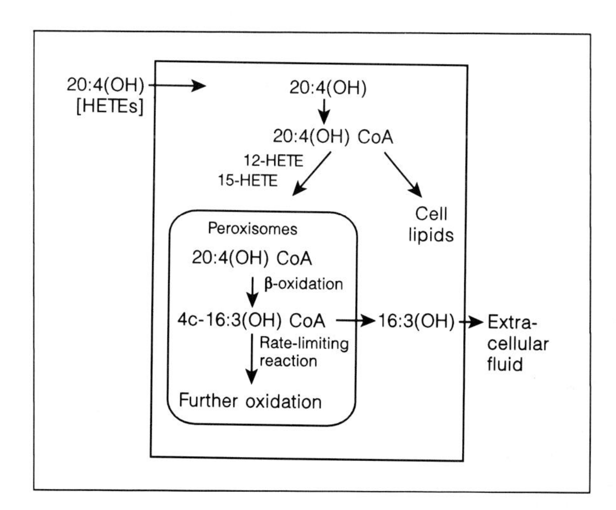

Fig. 3. Uptake and utilization of 12- and 15-HETE by mammalian cells.

slow step encountered in the peroxisomal β-oxidation pathway may be the 2,4-dienoyl CoA reductase reaction which is required for continued β-oxidation because the 16:3 metabolites contain 4-*cis* unsaturation.

As it builds up, some of the 16:3[OH] CoA is hydrolyzed and is released from the cell. This may be a mechanism for transporting these intermediates to tissue like the intestinal mucosa where they can rapidly undergo further oxidation [24]. Alternatively, it is possible that the formation of these intermediates is a specifically directed process because they have a functional role.

References

1 Spector AA, Gordon JA, Moore SA: Hydroxyeicosatetraenoic acids (HETEs). Prog Lipid Res 1988;27:271–323.

2 Moore SA, Giordano MJ, Kim H-Y, et al: Brain microvessels 12-hydroxyeicosatetraenoic acid is the (S) enantiomer and is lipoxygenase derived. J Neurochem 1991;57:922–929.

3 Stenson WF, Nickells MW, Atkinson JP: Esterification of monohydroxyfatty acids into the lipids of a macrophage cell line. Prostaglandins 1983;26:255–264.

4 Schafer AI, Takayama H, Farrell S, et al: Incorporation of platelet and leukocyte lipoxygenase metabolites by cultured vascular cells. Blood 1986;67:373–378.

5 Marcus AJ, Safier LB, Ullman HL, et al: 12S,20-dihydroxyicosatetraenoic acid: A new icosanoid synthesized by neutrophils from 12S-hydroxyicosatetraenoic acid produced by thrombin- or collagen-stimulated platelets. Proc Natl Acad Sci USA 1984;81:903–907.

6 Brinkman HJM, van Buul-Wortelboer MF, van Mourik JA: Selective conversion and esterification of monohydroxyeicosatetraenoic acids by human vascular smooth muscle cells: Relevance to smooth muscle cell proliferation. Exp Cell Res 1991;192:87–92.

7 Piomelli D, Greengard P: Lipoxygenase metabolites of arachidonic acid in neuronal transmembrane signalling. Trends Pharmacol Sci 1990;11:367–373.

8 Shen X-Y, Figard PH, Kaduce TL, et al: Conversion of 15-hydroxyeicosatetraenoic acid to 11-hydroxyhexadecatrienoic acid by endothelial cells. Biochemistry 1988;27:996–1004.

9 Brezinski ME, Serhan CN: Selective incorporation of (15S)-hydroxyeicosatetraenoic acid in phosphatidylinositol of human neutrophils: Agonist-induced deacylation and transformation of stored hydroxyeicosanoids. Proc Natl Acad Sci USA 1990;87:6248–6252.

10 Legrand AB, Lawson JA, Meyrick BO, et al: Substitution of 15-hydroxyeicosatetraenoic acid in the phosphoinositide signaling pathway. J Biol Chem 1991;266:7570–7577.

11 Hadjiagapiou C, Spector AA: 12-Hydroxyeicosatetraenoic acid reduces prostacyclin production by endothelial cells. Prostaglandins 1986;31:1135–1144.

12 Moore SA, Prokuski LJ, Figard PH, et al: Murine cerebral microvascular endothelium incorporate and metabolize 12-hydroxyeicosatetraenoic acid. J Cell Physiol 1988;137:75–85.

13 Gordon JA, Spector AA: Effects of 12-HETE on renal tubular epithelial cells. Am J Physiol 1987; 253:C277–C285.

14 Gordon EEI, Gordon JA, Spector AA: HETEs and coronary artery endothelial cells: Metabolic and functional interactions. Am J Physiol 1991;261:C623–C633.

15 Mathur SN, Field FJ, Spector AA, et al: Increased production of lipoxygenase products by cholesterol-rich mouse macrophages. Biochim Biophys Acta 1985;837:13–19.

16 Yla-Herttuala S, Rosenfeld ME, Parthasarathy S, et al: Colocalization of 15-lipoxygenase mRNA and protein with epitopes of oxidized low density lipoproteins in macrophage-rich areas of atherosclerotic lesions. Proc Natl Acad Sci USA 1990;87:6959–6963.

17 Wang L, Kaduce TL, Spector AA: Localization of 12-hydroxyeicosatetraenoic acid in endothelial cells. J Lipid Res 1990;31:2265–2276.

18 Gordon JA, Figard PH, Quinby GE, et al: 5-HETE: Uptake, distribution, and metabolism in MDCK cells. Am J Physiol 1989;256:C1–C10.

19 Gordon JA, Figard PH, Spector AA: Hydroxyeicosatetraenoic acid metabolism in cultured human skin fibroblasts: Evidence for peroxisomal β-oxidation. J Clin Invest 1990;85:1173–1181.

20 Mathur SN, Field FJ: Effect of cholesterol enrichment on 12-hydroxyeicosatetraenoic acid metabolism by mouse peritoneal macrophages. J Lipid Res 1987;28:1166–1176.
21 Hadjiagapiou C, Sprecher H, Kaduce TL, et al: Formation of 8-hydroxyhexadecatrienoic acid by vascular smooth muscle cells. Prostaglandins 1987;34:579–589.
22 Mathur SN, Albright E, Field FJ: 12-Hydroxyeicosatetraenoic acid is metabolized by β-oxidation in mouse peritoneal macrophages: Identification of products and proposed pathway. J Biol Chem 1990;265:21048–21055.
23 Hadjiagapiou C, Travers JB, Fertel RH, et al: Metabolism of 15-hydroxy-5,8,11,13-eicosatetraenoic acid by MOLT-4 cells and blood T-lymphocytes. J Biol Chem 1990;265:4369–4373.
24 Riehl TE, Bass NM, Stenson WF: Metabolism of 15-hydroxyeicosatetraenoic acid by CaCo-2 cells. J Lipid Res 1990;31:773–780.
25 Gordon JA, Zoeller RA, Spector AA: Hydroxyeicosatetraenoic acid oxidation in Chinese hamster ovary cells: A peroxisomal metabolic pathway. Biochim Biophys Acta 1991;1085:21–28.
26 Wigren J, Herbertsson H, Tollbom O, et al: Metabolism of 12(S)-hydroxy-5,8,10,14-eicosatetraenoic acid by kidney and liver peroxisomes. J Lipid Res 1993;34:625–631.
27 Dommes V, Baumgart C, Kunau W-H: Degradation of unsaturated fatty acids in peroxisomes: Existence of a 2,4-dienoyl-CoA reductase pathway. J Biol Chem 1981;256:8259–8262.

Dr. Arthur A. Spector, Department of Biochemistry, 4-403 BSB, University of Iowa, Iowa City, IA 52242 (USA)

Galli C, Simopoulos AP, Tremoli E (eds): Fatty Acids and Lipids: Biological Aspects.
World Rev Nutr Diet. Basel, Karger, 1994, vol 75, pp 16–17

Summary Statement: Fatty Acid Oxidation

Howard Sprecher

The session was co-chaired by *A. A. Spector* and *H. Sprecher,* and presentations were made by Drs. *Spector, Sprecher, H. Schultz, C. R. Roe,* and *S. Skrede.*

During the past several years the pathways for β-oxidizing unsaturated fatty acids have and continue to undergo revision. This is particularly true of acids with their first double bond at either position 4 or 5. For many years it was accepted that the removal of a double bond at position 4 proceeded as follows: 4-*cis*-acyl-CoA → 2-*trans*, 4-*cis*-acyl-CoA → → acetyl-CoA + 2-*cis*-acyl-CoA → D-β-hydroxyacyl-CoA → L-β-hydroxyacyl-CoA → continued B-oxidation. This pathway was shown to be incorrect with the discovery of 2-*trans*, 4-*cis*-dienoyl-CoA reductase. This enzyme catalyzes the NADPH-dependent reduction of the 2,4-conjugated diene to yield an acyl-CoA with its double bond at position 3. The double bond at position 3 is then isomerized to the 2-*trans* position by Δ-3-*cis* (trans)-Δ-2-enoyl-CoA isomerase to give an intermediate which is further β-oxidized. Indeed the 2,4-dienoyl-CoA reductase is a required enzyme for removing all double bonds at even numbered carbon atoms. According to this pathway, intermediates containing a 2-*cis* double bond would never be produced during β-oxidation. However, the presence of an epimerase which inverts the configuration of the hydroxyl group from the D- to the L-position has long been recognized. Recently this activity was shown to require two enzymes rather than a single protein. One of these enzymes is crotonase while the other is a hydratase that is specific for D-β-hydroxyacyl-CoA. The epimerization reaction may thus be depicted as follows: L-β-hydroxyacyl-CoA ↔ 2-*trans*-acyl-CoA ↔ D-β-hydroxyl-CoA. The role for D-hydroxyacyl-CoA derivatives in fatty acid β-oxidation is not as yet fully understood since only L-acyl-CoAs are substrates for B-hydroxyacyl-CoA dehydrogenase.

The β-oxidation of fatty acids with their first double bond at position 5 was generally thought to yield acetyl-CoA and a chain shortened intermediate with its first double bond at position 3. This pathway may also be incorrect since recent work has shown that the double bond at position 5 is removed by a so-called 5-reductase which in essence consists of the following rather complex set of reactions: 5-*cis*-acyl-CoA → 2-*trans*-5-*cis*-dienoyl-CoA → 3-*trans*-5-*cis*-dienoyl-CoA → 2-*trans*-4-*trans*-dienoyl-CoA → 3-enoyl-CoA → 2-*trans*-enoyl-CoA → β-oxidation. This reaction sequence requires the participation of a new enzyme that catalyzes the conversion of the 3-*trans*-5-*cis*-conjugated diene to one containing a 2-*trans*-4-*trans* configuration. It is not yet known whether this is the only pathway for removing double bonds at position 5 or whether the old pathway is also operative.

There are now many documented cases where fatty acid oxidation does not proceed to completion once the process is initiated. Several inborn errors of fatty acid oxidation have been described which are associated with the absence of an enzyme in this process. These defects may be associated with the transport of fatty acids into mitochondria or with the various enzymes that carry out reactions in the β-oxidation spiral. Most often these deficiencies are due to the lack of an acyl-CoA dehydrogenase. It has long been recognized that mitochondria contain short, medium and long chain acyl-CoA dehydrogenases which are matrix enzymes. Recently a very long chain acyl-CoA dehydrogenase has been shown to be associated with the inner mitochondrial membrane. Inborn errors of metabolism have been described which are due to the lack of all four enzymes; however, the most frequently encountered defect is due to the lack of medium chain length dehydrogenase.

Secondly, several laboratories have now shown that the lipoxygenase products, 15-hydroxy-5,8,11,13- and 12-hydroxy-5,8,10,14-eicosatetraenoic acids are only partially β-oxidized. The possibility thus exists that these chain-shortened metabolites might have biological activity or alternatively they may modify cellular processes by esterification into membrane lipids.

Finally, it has long been recognized that 22-carbon ω6 and ω3 fatty acids are partially β-oxidized with the resulting products being esterified into membrane lipids. This partial β-oxidation process may play a role in 22:6ω3 and 22:5ω6 biosynthesis since the production of these two acids requires intracellular communication between the endoplasmic reticulum and a site for β-oxidation which may be peroxisomal.

Galli C, Simopoulos AP, Tremoli E (eds): Fatty Acids and Lipids: Biological Aspects.
World Rev Nutr Diet. Basel, Karger, 1994, vol 75, pp 18–21

An Overview of the Pathways for the β-Oxidation of Polyunsaturated Fatty Acids

Horst Schulz

Department of Chemistry, City College, and Graduate School of the City University of New York, N.Y., USA

The degradation of unsaturated fatty acids, which generally contain *cis* double bonds, proceeds by β-oxidation and yields acetyl-CoA. This process requires the involvement of several auxiliary enzymes in addition to the enzymes of the β-oxidation spiral. Chain shortening of unsaturated fatty acids brings double bonds close to the CoA thioester function of fatty acyl-CoAs. At this stage, auxiliary enzymes act on double bonds to facilitate the complete β-oxidation of unsaturated fatty acids.

Even-Numbered Double Bonds

Unsaturated fatty acids with double bonds extending from even-numbered carbon atoms, e.g., the 12-*cis* double bond of linoleic acid, are chain shortened by β-oxidation to 4-*cis*-enoyl-CoAs (Cpd. 1; see figure 1 for the structure of compound 1 and for the structures of all other listed compounds), and then converted to 2-*trans*-4-*cis*-dienoyl-CoAs (Cpd. 2) by acyl-CoA dehydrogenase (EC 1.3.99.3) [1]. 2,4-Dienoyl-CoAs (Cpd. 2) are substrates of 2,4-dienoyl-CoA reductase (EC 1.3.1.34) which catalyzes the NADPH-dependent removal of one of the two double bonds and in mammals yields 3-*trans*-enoyl-CoAs (Cpd. 3) [1]. The latter compounds are acted upon by Δ^3, Δ^2-enoyl-CoA isomerase (EC 5.3.3.8) which isomerizes both the *cis* and *trans* isomers of 3-enoyl-CoAs to 2-*trans*-enoyl-CoAs (Cpd. 4) [1]. Since 2-*trans*-enoyl-CoAs are intermediates formed during the β-oxidation of saturated fatty acids, they are further degraded by the β-oxidation spiral. Overwhelming evidence supports the proposed degradation of unsaturated fatty acids with even-numbered double

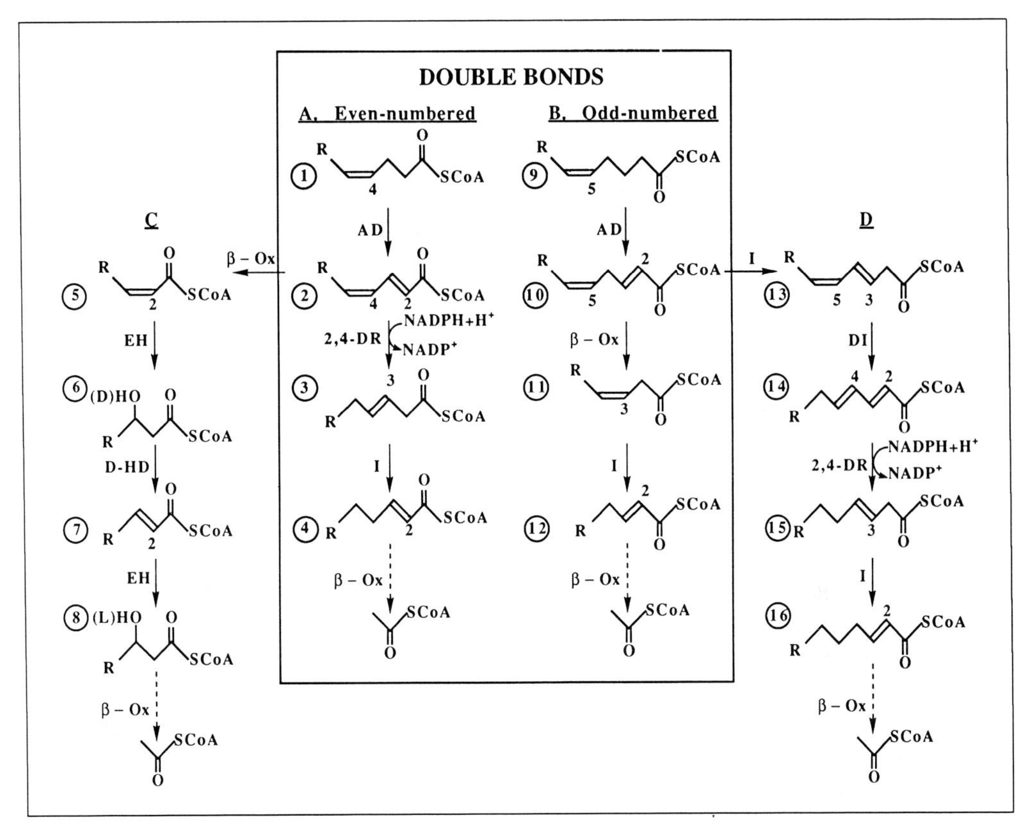

Fig. 1. β-Oxidation of unsaturated fatty acids with double bonds extending from even-numbered and odd-numbered carbon atoms. AD = Acyl-CoA dehydrogenase; 2,4-DR = 2,4-dienoyl-CoA reductase; I = Δ^3, Δ^2-enoyl-CoA isomerase; EH = enoyl-CoA hydratase; D-HD = D-3-hydroxyacyl-CoA dehydratase; DI = dienoyl-CoA isomerase; β-Ox = β-oxidation.

bonds by the reductase-dependent pathway (see fig. 1A). The evidence based on in vitro experiments is summarized and reviewed in Schulz and Kunau [1]. In vivo experiments also support this conclusion. For example, the absence of 2,4-dienoyl-CoA reductase from an *Escherichia coli* mutant prevented the metabolism 6-*cis*-octadecenoic acid (petroselinic acid) [2] and low levels of 2,4-dienoyl-CoA reductase in a patient with persistent hypotonia resulted in the excretion of 2-*trans*-4-*cis*-decadienoylcarnitine in the urine [3].

The proposed mitochondrial β-oxidation of unsaturated fatty acids with even-numbered double bonds by an epimerase-dependent pathway (see fig. 1C) is not supported by available evidence [1]. Most importantly, the chain shortening of 2-*trans*-4-*cis*-decadioenoyl-CoA (Cpd. 2) to 2-*cis*-enoyl-CoA (Cpd. 5) by β-oxidation does not proceed at a measurable rate in mitochondria [1]. Also, mitochondria do not contain 3-hydroxyacyl-CoA epimerase (EC 5.1.2.3) necessary for the conversion of D-3-hydroxyacyl-CoAs (Cpd. 6) to L-3-hydroxyacyl-CoAs (Cpd. 8) which are substrates of β-oxidation. Mammalian peroxisomes, however, contain a 3-hydroxyacyl-CoA epimerase activity [1] which is not a distinct enzyme, but is due to the combined actions of two enoyl-CoA hydratases with opposite stereospecificities [4]. D-3-Hydroxyacyl-CoAs (Cpd. 6) are converted by D-3-hydroxyacyl-CoA dehydratase to 2-*trans*-enoyl-CoAs (Cpd. 7) which are re-hydrated by enoyl-CoA hydratase (EC 4.2.1.17) to L-3-hydroxyacyl-CoAs.

Odd-Numbered Double Bonds

Chain shortening of unsaturated fatty acids with double bonds extending from odd-numbered carbon atoms, e.g., the 9-*cis* double bond of oleic acid, leads to the formation of 3-*cis*-enoyl-CoAs (Cpd. 11) which are converted by Δ^3,Δ^2-enoyl-CoA isomerase to 2-*trans*-enoyl-CoAs (Cpd. 12) [1]. The latter compounds are substrates of the β-oxidation spiral. A recent report claiming that the β-oxidation of 5-*cis*-enoyl-CoAs (Cpd. 9) by rat liver mitochondria requires NADPH [5] prompted a detailed study of the mitochondrial metabolism of 5-*cis*-octenoyl-CoA (Cpd. 9). This study [6] revealed that 2-*trans*-5-*cis*-octadienoyl-CoA (Cpd. 10), formed from 5-*cis*-octenoyl-CoA (Cpd. 9) by medium-chain acyl-CoA dehydrogenase, can be isomerized to 3,5-octadienoyl-CoA (Cpd. 13) by Δ^3,Δ^2-enoyl-CoA isomerase. 3,5-Octadienoyl-CoA undergoes further isomerization to 2-*trans*-4-*trans*-octadienoyl-CoA (Cpd. 14) by a novel enzyme, named dienoyl-CoA isomerase [M. J. Luo, et al., unpubl. results]. The resultant 2-*trans*-4-*trans*-octadienoyl-CoA can be reduced by NADPH-dependent 2,4-dienoyl-CoA reductase to 3-*trans*-octenoyl-CoA (Cpd. 15) which is isomerized to 2-*trans*-octenoyl-CoA (Cpd. 16) for further β-oxidation.

Conclusion

The β-oxidation of unsaturated fatty acids with even-numbered double bonds proceeds by the reductase-dependent pathway (fig. 1A). In peroxisomes, but not in mitochondria, the epimerase-dependent pathway (fig. 1C) may account for the β-oxidation of a small percentage (2%) of unsaturated fatty acids with even-numbered double bonds [1]. Unsaturated fatty acids with odd-numbered double bonds are assumed to be degraded by pathway B (see fig. 1) which only requires Δ^3,Δ^2-enoyl-CoA isomerase as an auxiliary enzyme. However, recently published studies suggest that odd-numbered double bonds may be reductively removed by NADPH-dependent 2,4-dienoyl-CoA reductase after two novel double bond shifts (see fig. 1D).

References

1 Schulz H, Kunau W-H: Beta-oxidation of unsaturated fatty acids: A revised pathway. Trends Biochem Sci 1987;12:403–406.
2 You S-Y, Cosloy S, Schulz H: Evidence for the essential function of 2,4-dienoyl-coenzyme A reductase in the β-oxidation of unsaturated fatty acids in vivo. J Biol Chem 1989;264:16489–16495.
3 Roe CR, Millington DS, Norwood DL, et al: 2,4-Dienoyl-coenzyme A reductase deficiency: A possible new disorder of fatty acid oxidation. J Clin Invest 1990;85:1703–1707.
4 Smeland TE, Li J, Cuebas D, et al: The mechanism and function of 3-hydroxyacyl-CoA epimerase in rat liver and *Escherichia coli;* in Coates PM, Tanaka K (eds): New Developments in Fatty Acid Oxidation. New York, Wiley-Liss, 1992, pp 85–93.
5 Tserng K-Y, Jin-S-Y: NADPH-dependent reductive metabolism of *cis*-5 unsaturated fatty acids. J Biol Chem 1991;266:11614–11620.
6 Smeland TE, Nada M, Cuebas D, et al: NADPH-dependent β-oxidation of unsaturated fatty acids with double bonds extending from odd-numbered carbon atoms. Proc Natl Acad Sci USA 1992; 89:6673–6677.

Dr. Horst Schulz, Department of Chemistry, City College of CUNY,
Convent Avenue at 138th Street, New York, NY 10031 (USA)

Galli C, Simopoulos AP, Tremoli E (eds): Fatty Acids and Lipids: Biological Aspects.
World Rev Nutr Diet. Basel, Karger, 1994, vol 75, pp 22–25

Inherited Defects of Fatty Acid Oxidation

Charles R. Roe, Mohamed A. Nada

Duke Medical Center, Durham, N.C., USA

Over the last decade many enzyme deficiencies in fatty acid oxidation have been identified and characterized [1]. These disorders are inherited in an autosomal recessive fashion and affect both saturated and unsaturated fatty acid degradation. For convenience, the disorders are divided into those of the carnitine cycle and those affecting intramitochondrial oxidation [2].

The carnitine cycle begins at the plasma membrane with the carnitine uptake system. Carnitine transverses the outer mitochondrial membrane where it is used by carnitine palmitoyl transferase I (CPT I) to form long-chain acylcarnitines. These products are transported via the carnitine/acylcarnitine translocase through the inner mitochondrial membrane where they become substrates for carnitine palmitoyl transferase II (CPT II). The product of this reaction is the long-chain acyl-CoA, the reactivated intermediate, which is further oxidized within the mitochondrion by the familiar process of β-oxidation. Inherited deficiencies of each of these steps in the carnitine cycle have been well characterized and are often fatal disorders.

Further oxidation of long-chain acyl-CoA compounds involves the well-known steps of β-oxidation, namely dehydrogenation catalyzed by acyl-CoA dehydrogenases, hydration which is catalyzed by enoyl-CoA hydratase, dehydrogenation which is catalyzed by 3-hydroxyacyl-CoA dehydrogenase and thiolysis catalyzed by thiolases resulting in the shortening of the longer chain substrate by producing acetyl-CoA. This four-step cycle then repeats for further degradation. This simplified description does not adequately explain the actual complexity of this metabolic pathway. Until recently, there were three acyl-CoA dehydrogenases: long-chain (LCAD), medium-chain (MCAD) and short-chain (SCAD) specific enzymes and the other enzymes of β-oxidation were considered to be in the mitochondrial matrix. It is now clear that there are additional systems in the inner mitochondrial membrane. An additional 'very long chain' acyl-CoA dehydrogenase (VLCAD) is now well characterized and inherited deficiencies of each of these four dehydrogenases are well established. Similarly, another inner membrane complex has been described which incorporates the activities of enoyl-CoA hydratase, L-3-hydroxyacyl-CoA dehydrogenase and thiolase. This membrane complex, trifunctional enzyme complex, appears to permit rapid channeling of the oxidative sequence for long-chain acyl-CoA compounds. Inherited deficiencies of these membrane components, VLCAD and trifunctional enzyme have also been recognized recently in children.

Further oxidation to the short-chain length appears to proceed mainly in the mitochondrial matrix. To date, no deficiencies have as yet been recognized for hydratase or thiolase in the matrix.

Unsaturated fatty acids with double bonds on even-numbered carbons require two additional enzymes as the chain length is shortened, namely 2,4-dienoyl-CoA reductase (DR) and enoyl-CoA isomerase. An inherited defect of DR has been described which validated the involvement of the reductase pathway in humans. As yet, no defect of enoyl-CoA isomerase has been identified. This enzyme is required for unsaturated fatty acids with double bonds on odd-numbered carbons. Similarly, the recently described dienoyl-CoA isomerase has not been associated with an inherited deficiency.

These disorders have been very difficult to recognize and diagnose. Specific enzyme assays from biopsied tissues by a few laboratories have been the primary means of identification. In some instances, metabolic analysis by gas chromatography combined with mass spectrometry (GC/MS) has been useful. More often these analyses demonstrate a problem with fat oxidation but not the specific enzyme defect, thereby requiring further investigation. The most powerful technique for simplified diagnosis of the majority of these inherited disorders is tandem mass spectrometry (MS-MS) analysis of whole blood or cultured fibroblasts (fig. 1). The result is an acylcarnitine profile which detects all acylcarnitine species from 2–18 carbon compounds from derivitized samples. This technology cannot recognize the carnitine uptake or CPT I deficiency since acylcarnitine intermediates do not accumulate in those disorders. However, from carnitine/acylcarnitine translocase, CPT II on down to SCAD deficiency, specific acylcarnitine profiles can be observed. Many of these disorders can present as early infant deaths without explanation. MCAD deficiency is associated with some deaths classified as sudden infant death syndrome. The tandem MS-MS technology can be used to analyze postmortem blood which was originally used for toxicology analysis and which is often stored up to 1 year after death. Similarly, a neonatal screening card, if available, can also be analyzed for these defects. Currently, automation of the tandem analysis for

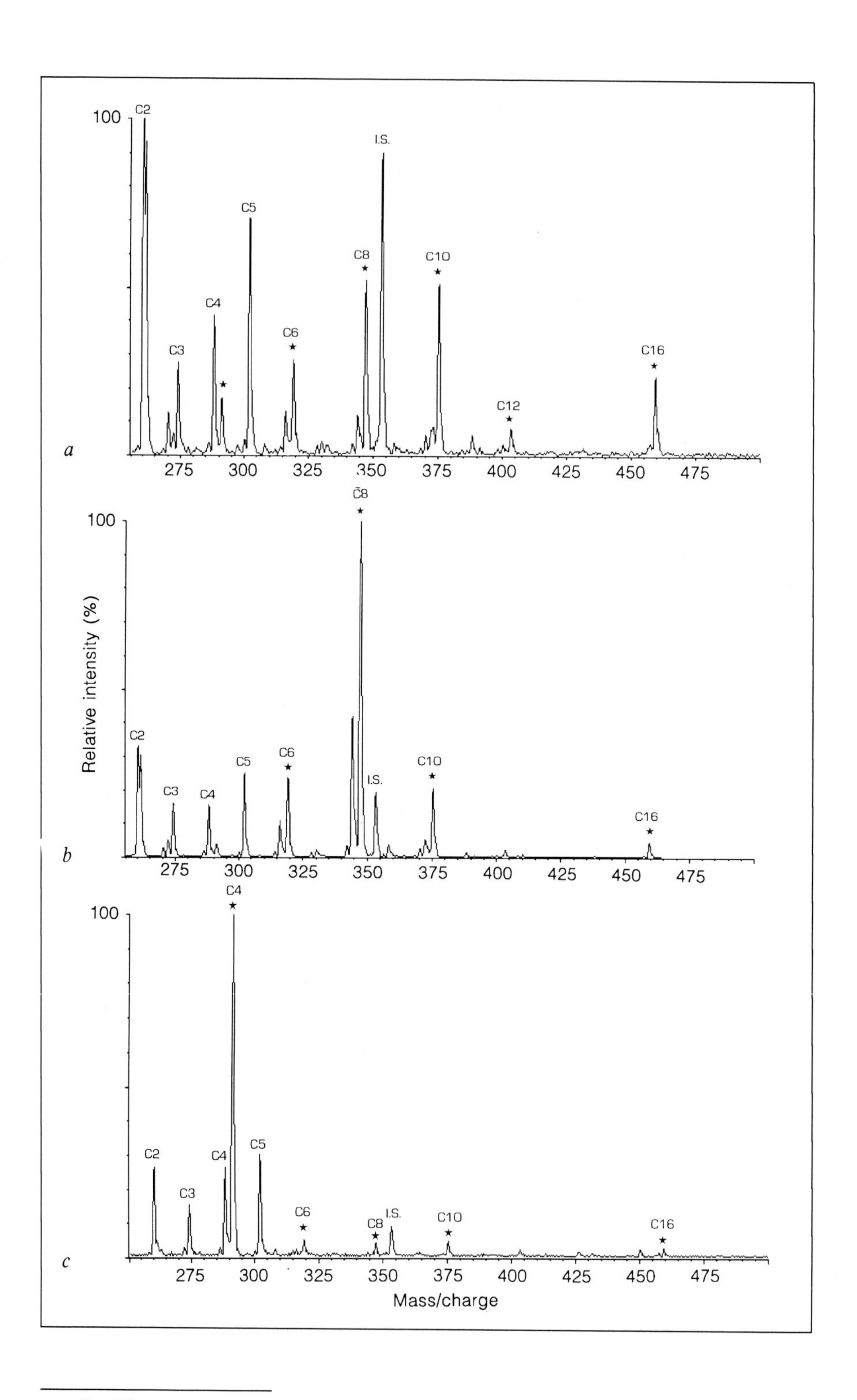
100
C2
I.S.
C5
C8
C10
C4
C6
C3
C16
C12
a
275
300
325
350
375
400
425
450
475
C8
100
Relative intensity (%)
C2
C5
C6
I.S.
C10
C3
C4
C16
b
275
300
325
350
375
400
425
450
475
C4
100
C2
C4
C5
C3
C6
C8
I.S.
C10
C16
c
275
300
325
350
375
400
425
450
475
Mass/charge

these and many other inherited disorders is being developed in an attempt to provide more sophisticated newborn screening and an opportunity to provide early treatment.

References

1 Roe CR, Coates PM: Mitochondrial fatty acid oxidation disorders; in Scriver CR, Beaudet AL, Sly WS, Valle D (eds): The Metabolic Basis of Inherited Disease, ed 7. New York, McGraw-Hill, 1994, in press.
2 Schultz H: Beta-oxidation of fatty acids. Biochim Biophys Acta 1991;1081:109–120.

Charles R. Roe, MD, Division of Biochemical Genetics, Department of Pediatrics, Duke Medical Center, PO Box 14991, Research Triangle Park, NC 27709 (USA)

Fig. 1. Acylcarnitine profiles of human skin fibroblasts incubated with *L*-carnitine and $^{2}H_{3}$-hexadecanoic acid. *a* Control. *b* MCAD-deficient. *c* SCAD-deficient. C_8 = Octanoylcarnitine; C_4 = butyrylcarnitine, etc.; IS = internal standard. Asterisk denotes labeled intermediates.

Galli C, Simopoulos AP, Tremoli E (eds): Fatty Acids and Lipids: Biological Aspects.
World Rev Nutr Diet. Basel, Karger, 1994, vol 75, pp 26–29

The Role Played by β-Oxidation in Unsaturated Fatty Acid Biosynthesis[1]

Howard Sprecher, Svetla Baykousheva

Department of Medical Biochemistry, The Ohio State University, Columbus, Ohio, USA

It has long been recognized that long-chain polyunsaturated fatty acids are partially β-oxidized with the subsequent esterification of chain-shortened intermediates into membrane lipids. For example, when 22:5ω6 [1] and 22:4ω6 [2] are fed to rats raised on a fat-free diet there is a large accumulation of 20:4ω6 in liver lipids. When [3-^{14}C]22:3ω6 was injected into rats fed chow it was possible to detect esterified 20:4 in microsomal and mitochondrial phospholipids suggesting that 22:3ω6 was chain shortened to 20:3ω6, at a site for β-oxidation, and then transferred to the endoplasmic reticulum where it was desaturated to yield 20:4ω6 [3]. The physiological relevance of this partial β-oxidation pathway remained undefined but the above studies were carried out before the discovery of peroxisomes. Recently we showed that 22:5ω3 was not desaturated to 22:6ω3 by a Δ 4-desaturase but that 22:5ω3 was metabolized in the endoplasmic reticulum as follows: 22:5ω3 → 24:5ω3 → 24:6ω3 [4]. The metabolism of 24:6ω3 to 22:6ω3 requires one revolution of the β-oxidation spiral followed by esterification of 22:6ω3 into membrane lipids. We have recently shown that 22:5ω6 is made via an analogous pathway thus requiring that 24:5ω6 is metabolized to 22:5ω6. It is generally accepted that peroxisomes partially β-oxidize long-chain fatty acids and that the chain-shortened compounds are transported, by some unknown process, to mitochondria where β-oxidation is completed [5]. We hypothesized that a role for peroxisomes might be to chain shorten fatty acids but that the product(s) would be transported to the endoplasmic reticulum where they would be esterified into lipids. The rationale to use [3-^{14}C]22:4ω6 as a substrate to evaluate this hypothesis stems

[1] Supported by NIH Grant DK 20387.

Table 1. Influence of microsomes and 1-acyl-*sn*-glycero-3-phosphocholine on the rate of peroxisomal β-oxidation of unsaturated fatty acids

Substrate	Minutes			
	A		B	
	5	30	5	30
[1-^{14}C]18:2ω6	28.7	57.7	14.1	23.5
[1-^{14}C]20:4ω6	9.0	31.1	5.0	11.4
[1-^{14}C]22:4ω6	25.8	69.1	23.9	69.7
[3-^{14}C]22:4ω6	3.8	19.1	2.3	5.0

Values are nanomoles of acid-soluble radioactivity produced when radiolabeled acids were incubated alone with peroxisomes(A) or along with microsomes and 1-acyl-*sn*-glycerol-3-phosphocholine(B).

from the observation that when it is intravenously injected into chow-fed rats, 96% of the esterified radioactivity in liver phospholipids is 20:4ω6 [6].

Peroxisomal Fatty Acid β-Oxidation

Highly purified peroxisomes were prepared by centrifuging the light mitochondrial fraction of liver from clofibrate-treated rats as described by Das et al. [7] except that we used a 35% Nycodenz gradient. Incubations were then carried out using 100 *μM* [1-^{14}C]18:2ω6, [1-^{14}C]20:4ω6, [1-^{14}C]22:4ω6 and [3-^{14}C]22:4ω6 with 300 μg of peroxisomal protein and the cofactors used by Bartlett et al. [8] when they showed that [U-^{14}C]16:0 was catabolized to give a homologous series of chain-shortened acyl-CoAs. The results in table 1 compare the nanomoles of acid-soluble radioactivity produced when incubation mixtures contained only peroxisomes as well as when they contained 300 μg of microsomal protein and 100 *μM* 1-acyl-*sn*-glycero-3-phosphocholine. Clearly, addition of microsomes and 1-acyl-*sn*-glycero-3-phosphocholine inhibits β-oxidation of all substrates except for [1-^{14}C]22:4ω6. After 30 min the lipids from all four incubation mixtures containing microsomes and 1-acyl-*sn*-glycero-3-phosphocholine were extracted by the procedure of Folch. With [1-^{14}C]18:2, [1-^{14}C]20:4, [1-^{14}C]22:4 and [3-^{14}C]22:4 the amount of radioactivity recovered in the organic phase for the four substrates was 56, 61, 14 and 62% respectively of what was added to the incubation. The upper aqueous layer contains both unreacted long-chain acyl-CoAs and acetyl-CoA formed by β-oxidation. When the lipids in the bottom layer were separated into neutral lipids and phospholipids, over 95% of the radioactivity was recovered in the phospholipid fraction. When the phospholipids from the incubation mixtures containing [1-^{14}C]18:2 and [1-^{14}C]20:4 were fractionated by thin-layer chromatography, 90 and 93% of the radioactivity respectively was associated with the choline phosphoglycerides. When [3-^{14}C]22:4 was the substrate, 78 and 15% of the radioactivity was esterified respectively into choline- and ethanolamine-phosphoglycerides. When the entire phospholipid fractions were interesterified and the methyl esters separated by reverse-phase HPLC, over 95% of the esterified radioactivity comigrated with 18:2ω6 and 20:4ω6 when [1-^{14}C]18:2 and [1-^{14}C]20:4 were the substrates. When [3-^{14}C]22:4 ω6 was the substrate, 75% of the esterified radioactivity was 20:4ω6 with the remainder being 22:4ω6. These data show that when microsomes and 1-acyl-*sn*-glycero-3-phosphocholine are added to peroxisomes that competing reactions take place – i.e. β-oxidation and esterification. With [1-^{14}C]18:2ω6 and [1-^{14}C]20:4ω6 the reduced rate of β-oxidation is due to esterification of the two substrates into an acceptor. The addition of microsomes and 1-acyl-*sn*-glycero-3-phosphocholine did not markedly alter the metabolism of 22:4ω6 to 20:4ω6 but once 20:4ω6 was produced it was preferentially esterified rather than β-oxidized.

Our findings raise several questions. We propose that β-oxidation of 24:6ω3 to 22:6ω3 and 24:5ω6 to 22:5ω6 are obligatory steps in providing 22:6ω3 and 22:5ω6 for membrane lipid biosynthesis. Why does partial β-oxidation of 24:5ω6 and 24:6ω3 at least in part stop when 22:5ω6 and 22:6ω3 are produced since 22:5ω6 [1] and 22:6ω3 [9] themselves are substrates for partial β-oxidation? Why does the nutritional status of the animal modify the activity of the partial β-oxidation-esterification process and why are there differences in the metabolism of ω3 and ω6 fatty acids [4, 6]? What role do mitochondria play in this partial degradative pathway? This is a particularly relevant question in light of the recent findings that mitochondria contain a fourth dehydrogenase which is specific for very-long-chain fatty acids with the resulting intermediates being channeled into a trifunctional enzyme to complete the β-oxidation spiral [10].

References

1 Verdino B, Blank ML, Privett OS, et al: Metabolism of 4,7,10,13,16-docosapentaenoic acid in the essential fatty acid-deficient rat. J Nutr 1964;83:234–238.
2 Sprecher H: The total synthesis and metabolism of 7,10,13,16-docosatetraenoate in the rat. Biochim Biophys Acta 1967;144:296–304.
3 Stoffel W, Ecker W, Assad H, et al: Enzymatic studies on the mechanism of the retroconversion of C22-polyenoic fatty acids to C20-homologues. Hoppe-Seylers Z Physiol Chem 1970;351:1545–1554.

4 Voss A, Reinhart M, Sankarappa S, et al: The metabolism of 7,10,13,16,19-docosapentaenoic acid to 4,7,10,13,16,19-docosahexaenoic acid is independent of a 4-desaturase. J Biol Chem 1991;266: 19995–20000.
5 Osmundsen H, Bremer J, Pedersen JI: Metabolic aspects of peroxisomal beta-oxidation. Biochim Biophys Acta 1991;1085:141–158.
6 Voss A, Reinhart M, Sankarappa S, et al: Difference in the interconversion between 20- and 22-carbon (n-3) and (n-6) polyunsaturated fatty acids in rat liver. Biochim Biophys Acta 1992;1127:33–40.
7 Das AK, Horie S, Hajra AK: Biosynthesis of glycerolipid precursors in rat liver peroxisomes and their transport and conversion to phosphatidate in the endoplasmic reticulum. J Biol Chem 1992; 267:9724–9730.
8 Bartlett K, Hovik R, Eaton S, et al: Intermediates of peroxisomal β-oxidation. Biochim J 1990; 270:175–180.
9 Fischer S, Vischer A, Preac-Mursic V, et al: Dietary docosahexaenoic acid is retroconverted in man to eicosapentaenoic acid, which can be quickly transformed to prostaglandin I_3. Prostaglandins 1987;34:367–375.
10 Uchida Y, Izai K, Orii T, et al: Novel fatty acid β-oxidation enzymes in rat liver mitochondria. J Biol Chem 1992;267:1034–1041.

Dr. Howard W. Sprecher, Department of Medical Biochemistry, The Ohio State University, 337 Hamilton Hall, 1645 Neil Avenue, Columbus, OH 43210 (USA)

Galli C, Simopoulos AP, Tremoli E (eds): Fatty Acids and Lipids: Biological Aspects.
World Rev Nutr Diet. Basel, Karger, 1994, vol 75, pp 30–34

Tetradecylthioacrylic Acid, a β-Oxidation Metabolite of Tetradecylthiopropionic Acid, Inhibits Fatty Acid Activation and Oxidation in Rat

Steinar Skrede, Pengfei Wu, Jon Bremer

Institute of Medical Biochemistry, University of Oslo, Norway

Tetradecylthiopropionic acid (TTP) is a stearic acid analogue with a sulfur atom substituting the methylene group in the 4-position of stearic acid (an alkyl-4-thiopropionic acid). TTP is metabolized by mitochondrial β-oxidation or by extramitochondrial ω-oxidation [1, 2]. In mitochondria, TTP is activated and rapidly oxidized to its 4-thia-*trans*-2-alkenoyl-CoA derivative, tetradecylthioacrylyl-CoA (TTAcr-CoA), which in turn is a poor substrate for further oxidation [1].

In contrast to the structurally closely related (non-β-oxidizable) 3-thia fatty acids, TTP does not lower serum lipids or induce peroxisomal proliferation in the rat [3]. Differently, rats fed TTP accumulated hepatic triacylglycerols [3]. In mitochondria isolated from rats treated with TTP, Hovik et al. [4] found reduced palmitoylcarnitine oxidation. Furthermore, TTP inhibited mitochondrial and peroxisomal β-oxidation in vitro, but its effector was not identified [4]. During oxidation of 4-thio-acylcarnitines, a progressive inhibition of β-oxidation was observed, leading to the assumption that a metabolite of the 4-thia-acylcarnitine was responsible for the reduced β-oxidation.

In the present study we have assumed that in rats fed TTP the inhibitor of β-oxidation per se is intramitochondrial TTAcr-CoA. We have now synthesized TTAcr (free acid) and investigated its effects on fatty acid metabolism in subcellular fractions of rat liver and in isolated rat hepatocytes.

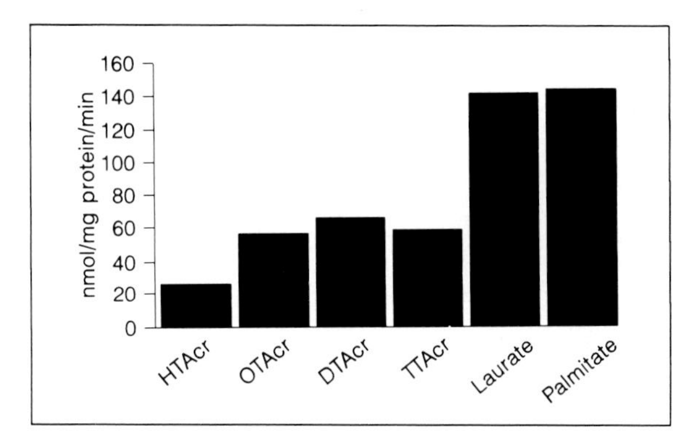

Fig. 1. Synthesis of acyl-CoA esters in rat liver microsomes (10 μg protein/0.4 ml) incubated with fatty acids (20 μ*M*), CoA (0.5 m*M*), ATP (5 m*M*) and $MgCl_2$ (5 m*M*) in Tris-HCl buffer (150 m*M*, pH 7.4) at 25 °C for 5 min. Formation of CoA esters from hexyl- (HTAcr), octyl- (OTAcr), decyl- (DTACr) and tetradecylthioacrylic acid (TTAcr) was measured spectrophotometrically at 312 nm, while formation of CoA esters from [1-^{14}C]lauric acid and [1-^{14}C]palmitic acid were used as references. The figure shows one representative experiment out of three.

Results

CoA Activation of Alkylthioacrylic Acids

Figure 1 shows that 4-thioacrylic acids of different chain lengths are activated to their CoA esters in rat liver microsomes. TTAcr was activated at a rate approximately 40% that of lauric acid or palmitic acid.

Acyl-[^{3}H]Carnitine Ester Formation in Mitochondria

Figure 2 shows the mitochondrial synthesis of acyl-[^{3}H]carnitines from carnitine and oleate, TTP or TTAcr. The carnitine ester formation from TTP took place at almost the same rate as that from oleic acid. TTAcr was a poor substrate for this reaction.

When synchronously adding oleic acid and 4-thia acid, TTP increased formation of acylcarnitines whereas TTAcr markedly reduced it (not shown). In incubations with purified carnitine palmitoyltransferase (CPT), TTAcr-CoA strongly inhibited palmitoyl-carnitine formation from palmitoyl-CoA (50% reduction at 0.6 μ*M* addition) (not shown).

With concomitant addition of oleic acid and 4-thia fatty acid, TTAcr inhibited mitochondrial oleic acid oxidation more than TTP at all concentrations. Differently, with a 2-min preincubation of mitochondria with the 4-thia fatty acids followed by an addition of [1-^{14}C]oleic acid, TTP inhibition was equal to that of TTAcr, which in turn was not increased by preincubation (not shown).

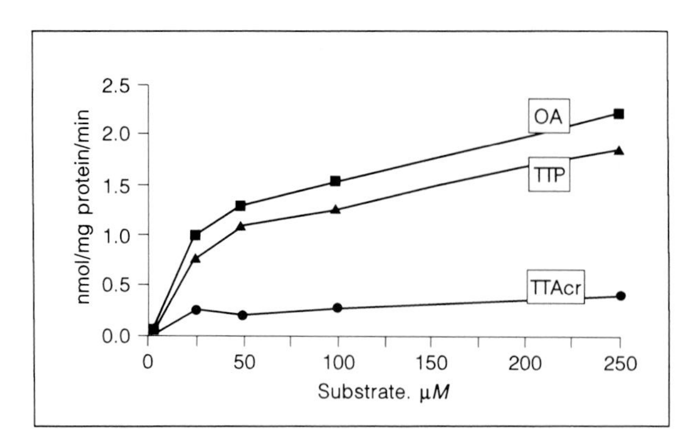

Fig. 2. Acylcarnitine formation from [methyl-^{3}H]carnitine (0.25 nM) and oleic acid, TTP or TTAcr in rat liver mitochondria (0.7 mg/1 ml) incubated with CoA (50 μM), ATP (5 mM), $MgCl_2$ (6 mM), BSA (0.4%), KCN (2 mM), MCl (80 mM) in Hepes buffer (25 mM, pH 7.3) for 6 min at 30 °C. The figure shows one typical experiment out of three.

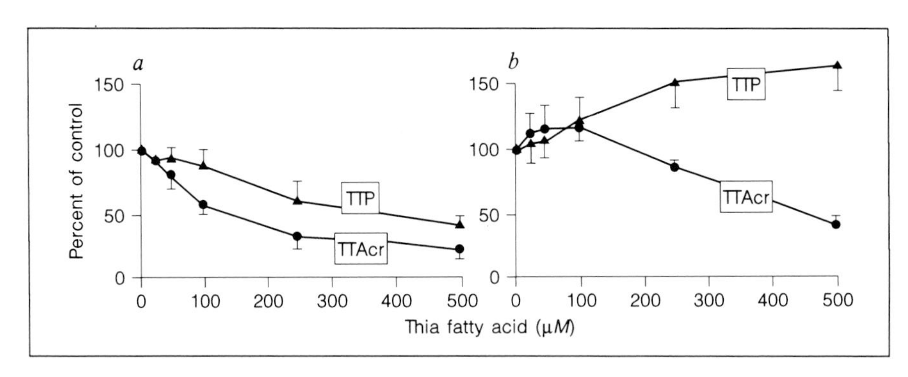

Fig. 3. Effects of TTP or TTAcr on the formation on (*a*) acid-soluble products or on (*b*) labelling of cellular lipids from [1-^{14}C]oleic acid (0.5 mM) by isolated hepatocytes (5 mg protein/ml) in KH buffer (pH 7.4) with 3.5% BSA incubated for 1 h. The results are given as means ± SD of 3–5 different cell preparations. Control rate of oxidation was 18.7 ± 3.4 nmol/mg protein/h, control rate of incorporation into lipids was 17.3 ± 3.7 nmol/mg protein/h.

Effects of 4-thia Fatty Acids on Rat Hepatocyte Metabolism of [1-^{14}C]Oleic Acid

Figure 3 shows that TTAcr inhibits the oxidation of oleic acid at all concentrations of TTAcr. Figure 3 also shows that at concentrations above 100 μM TTAcr inhibited incorporation of oleic acid into cellular lipids (in particular into triacylglycerols, not shown). At low concentrations of TTAcr, esterifica-

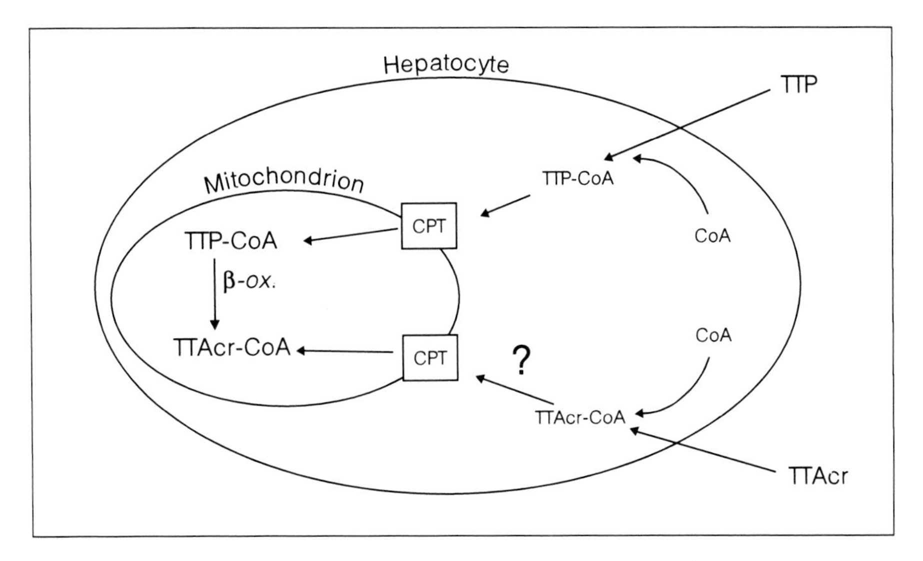

Fig. 4. Proposed scheme of effects of TTAcr TTP on hepatocyte lipid metabolism: TTAcr is activated outside mitochondria and exerts its effects extramitochondrially. TTP is activated outside mitochondria, translocated into mitochondria where it is β-oxidized and exerts its functions intramitochondrially, probably in the form of TTAcr-CoA.

tion was stimulated. Differently, when TTP inhibited oxidation (only at concentrations higher than 100 μ*M*) equimolar amounts of oleic acid were channeled into esterification reactions. Cellular uptake of oleic acid was consequently strongly reduced by TTAcr, whereas TTP had no influence on uptake.

Discussion and Conclusions

Effects of TTAcr on Rat Liver Metabolism

We have demonstrated that TTAcr is a substrate for CoA activation in endoplasmic reticulum, a reaction with large capacity in liver, but the CoA-ester formed is a poor substrate for CPT. On the contrary, TTAcr-CoA appears to be a potent inhibitor of the CPT reaction and fatty acid oxidation. TTAcr acylation of glycero-3-phosphate is also poor and TTAcr-CoA ester presumably inhibits also this enzyme (data not shown), as indicated by the reduced incorporation of oleic acid into lipids. In hepatocytes therefore, TTAcr exerts its functions mainly extramitochondrially (fig. 4). TTAcr-CoA formed outside mitochondria presumably accumulates, and fatty acid activation will be feed-back inhibited. This explains the reduced cellular uptake of oleic acid.

Effects of TTP on Rat Liver Metabolism

Differently, TTP is efficiently activated extramitochondrially and efficiently incorporated into lipids [2] or converted to carnitine ester for translocation into mitochondria [1, 2]. Intramitochondrial TTP is oxidized to TTAcr-CoA [1], which accumulates intramitochondrially and inhibits fatty acid oxidation, presumably at the level of acyl-CoA dehydrogenase.

References

1 Lau SN, Brantley RK, Thorpe C: 4-Thia-*trans*-2-alkenoyl-CoA derivatives: Properties and enzymatic reactions. Biochemistry 1989;28:8255–8262.

2 Hvattum E, Skrede S, Bremer J, et al: The metabolism of tetradecylthiopropionic acid, a 4-thia stearic acid, in the rat: In vivo and in vitro studies. Biochem J 1992;286:879–887.

3 Berge RK, Aarsland A, Kryvi H, et al: Alkylthio acetic acids (3-thia fatty acids) – A new group of non-β-oxidizable peroxisome-inducing fatty acid analogues. II. Dose-response studies on hepatic peroxisomal and mitochondrial changes and long chain fatty acid metabolizing enzymes in rats. Biochem Pharmacol 1989;38:3969–3979.

4 Hovik R, Osmundsen H, Berge RK, et al: Effects of thia-substituted fatty acids on mitochondrial and peroxisomal β-oxidation. Biochem J 1990;270:167–173.

Steinar Skrede, MD, Institute of Medical Biochemistry, University of Oslo, PO Box 1112, Blindern, N–0317 Oslo (Norway)

Galli C, Simopoulos AP, Tremoli E (eds): Fatty Acids and Lipids: Biological Aspects.
World Rev Nutr Diet. Basel, Karger, 1994, vol 75, pp 35–45

A New Look at Fatty Acids as Signal-Transducing Molecules

Gérard P. Ailhaud [a], *Nada Abumrad* [b], *Ez-Zoubir Amri* [a], *Paul A. Grimaldi* [a, 1]

[a] Centre de Biochimie (UMR 134 CNRS), Université de Nice-Sophia Antipolis, Faculté des Sciences, Nice, France;
[b] Department of Physiology and Biophysics, Health Sciences Center, SUNY, Stony Brook, N.Y., USA

Since the identification of long-chain fatty acids (FA) as essential components of cellular lipids, researchers have attempted to define their multiple properties and functions. FA were first characterized as quantitatively the most important substrates for phospholipids which are essential membrane components and of prostaglandins which have a variety of regulatory effects. More recently, FA were shown to directly regulate a range of biological processes. For example, FA modulate ion channel activation, enzyme function and synaptic transmission (reviewed by Ordway et al. [1]). This review does not attempt to summarize the many biological effects of FA. It only addresses two areas of FA research where significant new insights have been gained over the last few years. These areas relate to the mechanism of *FA entry into cells* and to the role of intracellular *FA as transcriptional regulators.* The central aim will be to highlight the importance of FA supply to the cell in regulating expression of lipid-related genes as this might help to understand FA involvement in the pathophysiology of multiple conditions.

[1] The authors are grateful to Dr. F. McKenzie for careful reading of the manuscript, to Mrs. G. Oillaux for secretarial assistance and to C. Cibré and J. Sayegh for expert photographical work.

Membrane Transport of Long-Chain FA and Its Relationship to Cell Differentiation

The mechanism of FA transfer across cell membranes has long been postulated to occur by simple diffusion through the lipid bilayer. However, biophysical data with model lipid membranes has led to conflicting interpretations with respect to whether FA transfer is fast enough to account for cellular uptake. There is disagreement on the actual rate for transfer of native long-chain FA with reported values differing by more than 10^4 [2–5]. On the other hand, a large body of biochemical evidence has argued against FA transfer occurring mainly through membrane lipid. In rat adipocytes [6–8] this included saturability of transfer under conditions where the FA was not metabolized, sensitivity to protein-modifying agents and regulation by hormones. Features of FA transfer consistent with carrier mediation were also demonstrated in other cell types, notably hepatocytes [9]. More recently, large increases in FA transport have been documented during the differentiation of three different lines of preadipocytes in culture [10–12]. These precursor cells (adipoblasts) differentiate via multiple steps into adipose cells showing most of the biochemical and morphological characteristics of isolated adipocytes [13]. Activation for the gene of the adipocyte lipid-binding protein (ALBP), a well-studied differentiation marker, is observed at the beginning of triacylglycerol accumulation and following the emergence of multiple early markers (lipoprotein lipase, pOb24/A2COL6) [14] and clone 9 encoding for a protein of the pentaxin family [15, 16] [Amri et al., unpubl. observations]. The increase in FA transport was shown to occur at an early stage preceding the induction of ALBP and of FA synthetase and the beginning of triacylglycerol accumulation [10]. These findings raised the possibility that early increases in FA transport and thus in intracellular FA could modulate subsequent differentiation steps. The recently demonstrated effects of FA on expression of lipid-related genes, discussed in later sections, are consistent with this interpretation.

All of the above data support *protein-mediation of FA uptake* in mammalian cells. Although the identity of the putative FA carrier remains undetermined, significant strides have been made. Two proteins of different molecular weights have been implicated in the transport process. Previously reported findings with the FA-transport inhibitor [^{3}H]DIDS [7] suggested involvement of an 85- to 90-kD protein in the transfer process. More recent studies using a synthetic sulfosuccinimidyl derivative of oleate [17, 18] lend further support to this interpretation. The protein (88 kD) has been isolated and its amino-terminal sequence was determined [18]. Based on this sequence, a full coding clone (FAT88) was isolated recently from a fat cell cDNA library which showed considerable homology with that of human CD36 [20]. A screening of RNA from Ob1771 cells at different stages in culture with the FAT88 clone demonstrated the differentiation-dependent emergence of two mRNAs of about 3.0 and 4.8 kb. The mRNAs were up-regulated by exogeneous FA and by dexamethasone. Tissue distribution of the mRNA indicated its presence in heart, muscle, intestine, spleen and adipose tissues and its absence from liver and kidney. The recent availability of mouse fibroblast cells showing very limited ability to transport fatty acids should provide a valuable tool for determining the role of the 88-kD protein in the process [Grimaldi et al., unpubl. observations].

A 40-kD protein, isolated from hepatocytes, has also been implicated in FA transport in multiple cell types based on two lines of evidence. First, the protein bound to an oleate-agarose affinity column and, second, antibody raised against the protein inhibited oleate transport by 50–60% [21]. The 40-kD protein was recently reported to be similar in sequence and properties to the mitochondrial enzyme glutamate-oxaloacetate transaminase (mGOT) [22]. Expression of the protein increased during differentiation of 3T3-L1 cells and this system might be useful in future work aimed at cloning and establishing its role in the transport process [11].

FA Uptake and Expression of ALBP Are Differently Regulated in Adipose Cells

The contribution of FA-binding proteins (FABPs) to the translocation of FA across plasma membrane has remained a controversial issue [23]. Work with BFC-1 [24] and OB1771 [25] cell lines has offered unique opportunities to shed some light on this point. In general the data showed no correlation between activity of *FA uptake* and *expression of ALBP*. In differentiating BFC-1 cells, oleate uptake increases (5-fold) about 4 days before induction of ALBP mRNA [10]. In Ob1771 cells (fig. 1), basal FA uptake (0.14 nmol/min/10^6 cells) is increased 3- and 4-fold in cells treated for 24 and 48 h with dexamethasone (0.42 and 0.52 nmol/min/10^6 cells, respectively). In contrast, no significant induction of ALBP is observed under these conditions. In Ob1754 cells, a polyamine-dependent variant of Ob17 cells which requires growth hormone (GH) for terminal differentiation [26], the rate of FA uptake is not modified by the presence of either putrescine or GH. This is in contrast to the expression of the ALBP gene which is dependent on putrescine treatment (fig. 2). Taken together, these results indicate that FA uptake and ALBP gene expression are regulated differently and regulatory effects on the two parameters can be dissociated. The data do not rule out the possibility that another FABP has been induced by differentiation in a manner which correlates better with the

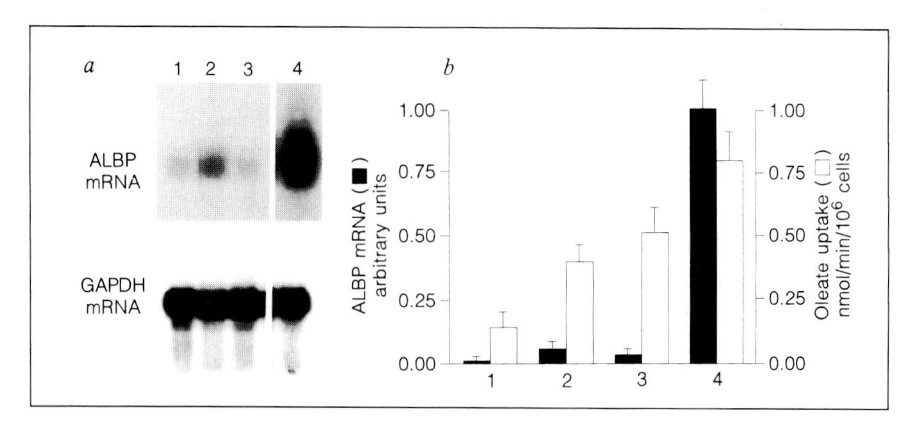

Fig. 1. Dissociation between FA uptake and ALBP expression in Ob1771 cells. One day after confluence, at which time the transition adipoblasts ⇒ preadipocytes takes place, Ob1771 cells were exposed in the differentiation medium to 1 μ*M* dexamethasone for 24 h (2) or 48 h (3). Untreated cells were assayed at 1 day (1) and 12 days (4) after confluence. Oleate uptake [10] and mRNA quantitation [27] were determined as previously described.

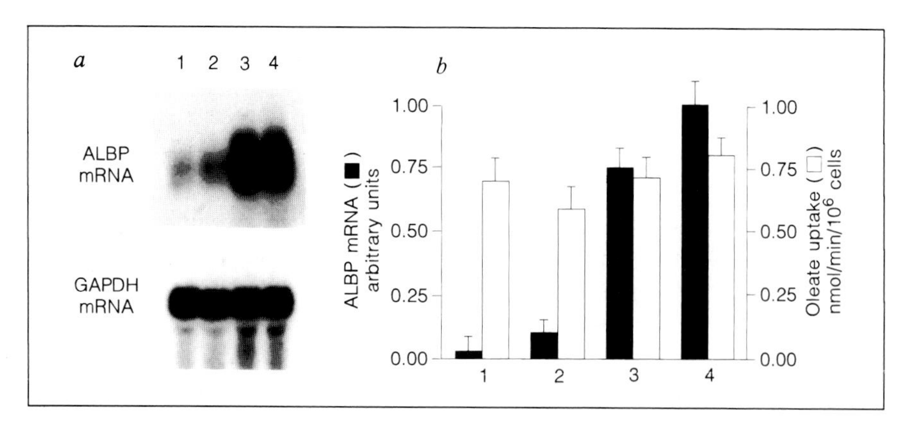

Fig. 2. Dissociation between FA uptake and ALBP expression in polyamine-dependent Ob1754 cells. Ob1754 cells were maintained in the differentiation medium 14 days after confluence in the absence (1) or in the presence of 2 n*M* growth hormone (2), 100 μ*M* putrescine (3), or both (4).

changes in FA uptake. However, no evidence exists for such a protein in adipose cells.

FA per se Are Inducers of the Expression of Lipid-Related Genes

Activation of the ALBP gene was first reported in committed, early marker-expressing, triacylglycerol-free Ob1771 cells where long-chain FA ($\geq C_{12}$, saturated and unsaturated) were found effective at micromolar concentrations. Once the ALBP gene is expressed, a reversible modulation of its expression by fatty acids was also observed in differentiated, triacylglycerol-containing Ob1771 cells [27, 28]. Similar but less pronounced changes were described in BFC-1 cells [10]. Although the total concentrations of FA added in these experiments tended to be high (300 μ*M*), the effective concentrations were likely to be low. The fraction of FA in solution which is not bound to serum albumin (unbound FA) is the fraction taken up by cells [29] and the one physiologically active. This fraction is determined by the molar ratios of fatty acid to albumin. The ratios used ranged from 0.8 to 5.7 [28, 29] so the corresponding unbound FA ranged from 40 n*M* to 5 μ*M*. Finally, medium FA are rapidly depleted by cells so the addition of high FA concentrations was necessary to keep a steady effective concentration of medium FA [29]. Consistent with this, α-bromopalmitate, a nonmetabolized long-chain FA, was much more potent than native FA in inducing expression of the ALBP gene in Ob1771 cells. The effective concentrations of α-bromopalmitate were about one fifth those for native F4 [29]. Interestingly, although early marker-expressing Ob1771 cells do not incorporate α-bromopalmitate into lipids, due to the absence of palmitoyl-CoA synthetase, its cellular uptake can be demonstrated: at a total extracellular concentration of 40 μ*M* labeled bromopalmitate and a FA serum albumin ratio of 4.0, the intracellular concentration of labeled bromopalmitate was about ~3 μ*M*. This value might relate to the intracellularly active FA concentration. It also approximated that of unbound palmitate in the culture medium under identical conditions, suggesting that bromopalmitate binds to serum albumin with an affinity similar to that of palmitate. When adipose precursor cells are fully differentiated, the intracellular ALBP concentration in the cytosol can be estimated within the millimolar range. The fraction of cell FA that is bound to ALBP is more difficult to estimate as a result of compartimentation of cell FA. The metabolically relevant pool of intracellular free FA is a very small percent of cell FA and generally does not exceed a few micromoles. This pool might be in equilibrium with ALBP-bound FA; however, it was shown to follow changes in exogenous unbound FA [30]. Thus low levels of exogeneous unbound FA can chronically up-regulate ALBP gene expression. *The regulation of ALBP gene* takes place primarily at a *transcriptional level* [28, and Grimaldi, unpubl. observations]; this transcriptional effect was not observed in 3T3-F442A cells [31], likely due to different experimental conditions.

Besides their effect on the ALBP gene, long-chain FA (saturated and unsaturated) activate the expression of an acyl-CoA synthetase (ACS) gene [28] which encodes an enzyme able to catalyze the conversion of palmitate and oleate to CoA derivatives [29]. This enzyme is induced by cell differentiation and appears important for the initiation of lipid deposition [29]. It allows the cell to metabolize efficiently a wide range of long-chain FA substrates which include those preferentially incorporated into triglycerides. The effect of FA on this enzyme, similar to that on ALBP, is consistent with intracellular FA acting to promote differentiation of preadipocytes. However, the effect of FA on ACS was not specific to adipocytes since activation of the gene by α-bromopalmitate supplementation was also observed in rat FA32 hepatoma cells [K. Schoonjans et al., unpubl. observations]. In conclusion, it is proposed that, both in preadipocytes and adipocytes, a *low intracellular concentration of unbound long-chain FA* (saturated and unsaturated) is able to modulate the expression of ALBP and ACS genes, and possibly that of other lipid-related genes. These effects might be exerted in a way similar to the effects of retinoic acid, a FA analogue which functions by binding and activating specific retinoic acid receptors [32]. It is also possible that some agents could exert their effects on gene expression indirectly via modulating intracellular concentration of unbound FA. For example, insulin, via its receptor kinase activity, has been reported to regulate the phosphorylation of ALBP, a process which dramatically reduces the affinity of ALBP for FA (from an apparent K_d of $\sim 1\ \mu M$ to $> 50\ \mu M$) [33]. This phosphorylation, which takes place at a low degree in intact cells (only 0.2% of intracellular pool ALBP becomes phosphorylated), has been postulated to target the protein to the microsomal membrane [33]. However, it could also promote release of FA from ALBP, increasing intracellular concentration of unbound FA and subsequently up-regulating expression of ALBP and other lipid-related genes.

New Members of the Steroid Hormone Receptor Superfamily Can Confer FA Responsiveness

Among the family of peroxisome proliferators (PPs), the most potent compounds in rodents are fibrates which, like FA, are *amphipathic carboxylates*. Recently, a member of the steroid horomone receptor superfamily which can be activated by PPs (i.e. *PP-activated receptor* PPAR) has been cloned and sequenced from a cDNA library of mouse liver and termed mPPARα [34].

Other PPARs have been identified from cDNA libraries of rat liver (rPPARα) [35], *Xenopus laevis* (xPPARα, β and γ) [36] and human osteosarcoma SAOS-2/B10 cells (hNUC1) [37]. PPARs, with molecular weight ranging from 45 to 54 kD, have been shown to bind to specific DNA sequences (termed the PP response element, PPRE) which have been located in the PP-responsive genes encoding acyl-CoA oxidase [38, 39], enoyl-CoA hydratase/3-hydroxyacyl-CoA dehydrogenase [40], cytochrome P-450 monooxygenase 4A6 [41], and possibly rat liver fatty acid-binding protein [42]. Evidence has been obtained that both PPs and long-chain FA (mainly unsaturated) can activate a chimera consisting of the ligand-binding domain of rPPARα and the DNA-binding domain of the human glucocorticoid receptor. Arachidonic and linoleic acids were shown effective in CHO cells stably expressing the chimeric receptor [35]. Similar findings were obtained with transient transfection of COS cells with a chimeric receptor composed of the ligand-binding domain of hNUC1 and the DNA-binding domain of the mouse glucocorticoid receptor: in this case, both arachidonic and oleic acids were effective [37]. Recently, Hepa1 hepatoma cells have been transiently cotransfected with an mPPARα expression vector and a reporter plasmid. The latter contains the PPRE regulatory sequences from rat peroxisomal acyl-CoA oxidase upstream of the rabbit β-globin promoter and the chloramphenicol acyltransferase (CAT) coding sequence. Both the potent peroxisome proliferator Wy-14,634 and palmitic acid activate mPPARα and lead to an increase in CAT activity [42]. Moreover, the synthetic arachidonic analogue 5,8,11,14-eicosatetraynoic acid is two orders of magnitude more potent than Wy-14,634 in the activation of xPPARα [43]. Clearly, all the PPARs so far described are activated by a diversity of molecules that include several amphipathic carboxylates. However, given the diversity of their chemical structure of active molecules and the failure to demonstrate any direct binding, one questions whether PPs are the natural ligands of PPARs. Stimulation of PPARs by FA occurs at concentrations much higher than those required for PPs. However, as already shown for adipocytes, this is likely to be due to their rapid metabolism and to the fact that the active concentration relates to that of unbound FA within the cell and this concentration is very low [30]. Alternatively, it is possible that PPAR-related proteins with high affinity for FA could mediate FA effects at the gene level. This interpretation would be consistent with the absence of mPPARα in mouse Ob1771 adipose cells, where FA can regulate expression of the genes for ALBP, ACS and lipoprotein lipase and also with the fact that mPPARα was not detected in FA-, α-bromopalmitate- or fibrate-treated Ob1771 cells. Quite recently, we have cloned and sequenced, from a cDNA library of FA-treated Ob1771 cells, a new member of the superfamily of steroid hormone receptor. It shows 70% homology with members of the PPAR family and was termed *FA-activated receptor* (FAAR) as

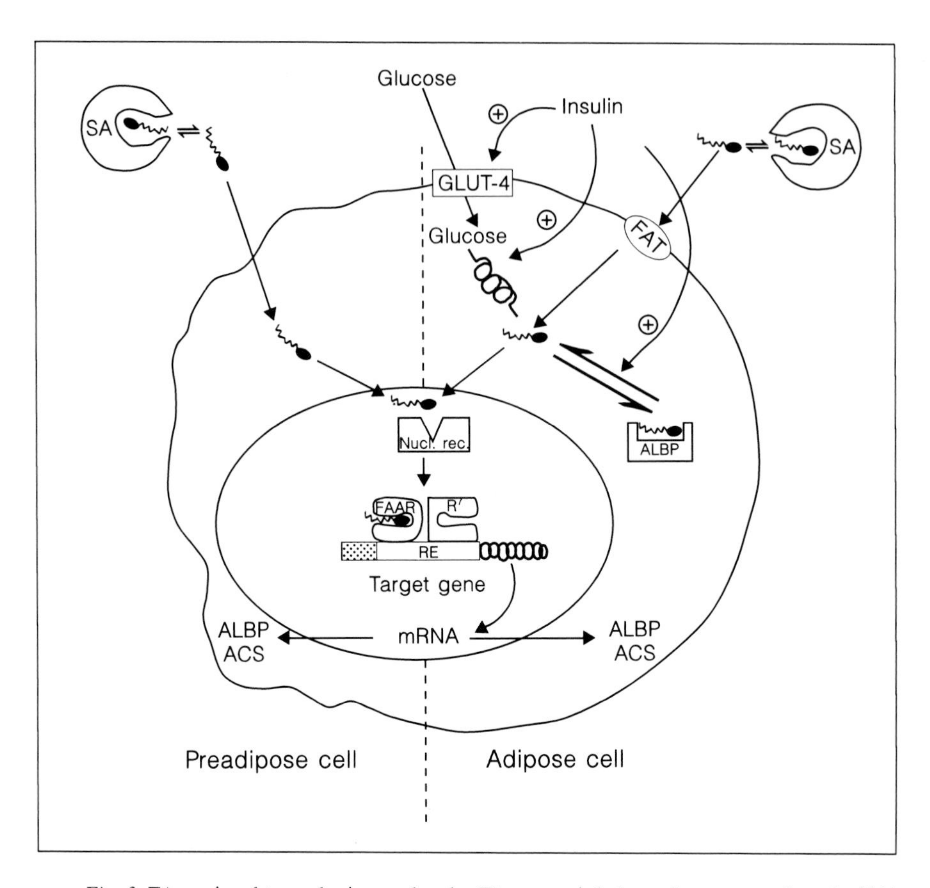

Fig. 3. FA as signal-transducing molecules FA are mainly bound to serum albumin (SA) but only unbound FA enter target cells. After entry into preadipose cells lacking ALBP (left half), FA of exogenous origin (lack of de novo fatty acid synthesis) can bind directly to FAAR. After inducing a conformational change of the latter, the active complex likely presents as heterodimer (see text) and interacts with target genes by binding to response elements; this interaction leads to the expression of ALBP and ACS genes, the synthesis of their cognate mRNAs and a subsequent increase in ALBP content. In adipose cells (right half), exogenous FA enter the cells by a protein-mediated process (FA transporter, FAT), whereas endogenous FA arise from glucose metabolism. When the FA supply is increased from either source, this leads to a rise in the concentration of FA not bound to ALBP which is sufficient to induce within the nucleus the formation of an active heterodimer. The modulation of ALBP gene expression is fully reversible upon removal of exogenous FA or following a decrease in de novo FA synthesis. Insulin stimulation takes place at the level of the glucose transporter GLUT-4, the FA synthesis pathway and the increased dissociation of FA from ALBP.

its expression is strongly and positively modulated by long-chain FA. Recent evidence indicates that PPARs are operating via association to other members of the *steroid hormone receptor superfamily* [43–45]: PPARs heterodimerize with retinoid X receptor and the cooperativity takes place through PPRE characterized in various PP-responsive genes [38–41, 46]. The possibility that FAAR operates in a similar manner remains to be investigated.

Conclusions

Our current working hypothesis, summarized in figure 3, stresses the importance of *FA as signal-transducing molecules*. Future work will address the mechanisms by which FA, independently of hormones, might mediate some of the chronic effects of diets or metabolic conditions on the expression of key functional proteins.

References

1 Ordway RN, Singer JJ, Walsh JV: Direct regulation of ion channels by fatty acids. Trends Neurosci 1991;14:96–100.
2 Gutknecht J: Proton conductance caused by long-chain fatty acids in phospholipid bilayer membranes. J Membr Biol 1988;106:83–93.
3 Kamp F, Hamilton JA: pH gradients across phospholipid membranes caused by fast flip-flop of un-ionized fatty acids. Proc Natl Acad Sci USA 1992;89:11367–11370.
4 Noy N, Donelly TM, Zakim D: Physical-chemical model for the entry of water-insoluble compounds into cells. Studies of fatty acid uptake by the liver. Biochemistry 1986;25:2013–2021.
5 Kleinfeld AM: Transport of free fatty acids across membranes. Comments. Mol Cell Biophys 1990;6:361–383.
6 Abumrad NA, Perkins RC, Park JH, et al: Mechanisms of long chain fatty acid permeation in the isolated adipocyte. J Biol Chem 1991;256:9183–9190.
7 Abumrad NA, Park JH, Park CR: Permeation of long-chain fatty acids into adipocytes. J Biol Chem 1984;259:8945–8953.
8 Abumrad NA, Park CR, Whitesell RR: Catecholamine activation of the membrane transport of long chain fatty acids in adipocytes is mediated by cyclic AMP and protein kinase. J Biol Chem 1986;261:13082–13089.
9 Stremmel W, Berk PD: Hepatocellular influx of [^{14}C]oleate reflects membrane transport rather than intracellular metabolism or binding. Proc Natl Acad Sci USA 1986;83:3086–3090.
10 Abumrad NA, Forest CD, Regen DM, et al: Increase in membrane uptake of long-chain fatty acids early during preadipocyte differentiation. Proc Natl Acad Sci USA 1991;88: 6008–6012.
11 Zhou SL, Stump D, Sorrentino D, et al: Adipocyte differentiation of 3T3-L1 cells involves augmented expression of a 43-kD plasma membrane fatty acid-binding protein. J Biol Chem 1992; 267:14456–14461.
12 Trigatti BL, Mangroo D, Gerber GE: Photoaffinity labelling and fatty acid permeation in 3T3-L1 adipocytes. J Biol Chem 1991;266:22621–22625.
13 Ailhaud G, Grimaldi P, Négrel R: Cellular and molecular aspects of adipose tissue development. Annu Rev Nutr 1992;12:207–233.
14 Ibrahimi A, Bertrand B, Bardon S, et al: Cloning of α2 chain of type VI collagen and expression during mouse development. Biochem J 1993;289:141–147.
15 Lee GW, Lee TH, Wilcek J: TSG-14, a tumor necrosis factor- and IL-1-inducible protein, is a novel member of pentaxin family of acute phase proteins. J Immunol 1993;150: 1804–1812.

16 Breviario F, d'Aniello EM, Golay J, et al: Interleukin-1-inducible genes in endothelial cells: Cloning of a new gene related to C-reactive protein and serum amyloid P component. J Biol Chem 1992;267:22190–22197.
17 Harmon CM, Luce P, Beth AH, et al: Labeling of adipocyte membranes by sulfo-n-succinimidyl derivatives of long-chain fatty acids: Inhibition of fatty acid transport. J Membr Biol 1991; 121:261–268.
18 Harmon CM, Luce P, Abumrad NA: Labelling of an 88-kD adipocyte membrane protein by sulpho-n-succinimidyl long-chain fatty acids: Inhibition of fatty acid transport. Biochem Soc Trans 1992;20:811–813.
19 Harman CM, Abumrad NA: Binding of sulfosuccinimidyl fatty acids to adipocyte membrane proteins. Isolation and amino-terminal sequence of an 88-kD protein implicated in transport of long-chain fatty acids. J Membr Biol 1993;133:43–49.
20 Abumrad NA, Raafat El-Maghrabi M, Amri EZ, et al: Cloning of a rat adipocyte membrane protein implicated in binding or transport of long-chain fatty acids that is induced during preadipocyte differentiation. J Biol Chem 1993;268:17665–17668.
21 Potter BJ, Sorrentino D, Berk PD: Mechanisms of cellular uptake of free fatty acids. Annu Rev Nutr 1989;9:253–270.
22 Berk PD, Wada H, Horio Y, Potter B, et al: Plasma membrane fatty acid-binding protein and mitochondrial glutamic-oxaloacetic transaminase of rat liver are related. Proc Natl Acad Sci USA 1990;87:3484–3488.
23 Glatz JFC, van der Vusse GJ: Cellular fatty acid-binding proteins: Current concepts and future directions. Mol Cell Biochem 1990;98:237–251.
24 Forest C, Doglio A, Ricquier D, et al: A preadipocyte clonal line from mouse brown adipose tissue. Short- and long-term responses to insulin and β-adrenergics. Exp Cell Res 1987;168:218–232.
25 Doglio A, Dani C, Grimaldi P, et al: Growth hormone regulation of the expression of differentiation-dependent genes in preadipocyte Ob1771 cells. Biochem J 1986;238:123–129.
26 Amri E, Dani C, Doglio A, et al: Adipose cell differentiation. Evidence for a two-step process in the polyamine-dependent Ob1754 clonal line. Biochem J 1986;238:115–122.
27 Amri E, Bertrand B, Ailhaud G, et al: Regulation of adipose cell differentiation. I. Fatty acids are inducers of the aP2 gene expression. J Lipid Res. 1991;32:1449–1456.
28 Amri E, Bertrand B, Ailhaud G, et al: Regulation of adipose cell differentiation. II. Kinetics of induction of the aP2 gene by fatty acids and modulation of dexamethasone. J Lipid Res 1991;32: 1457–1463.
29 Grimaldi PA, Knobel SM, Whitesell RR, et al: Induction of the aP2 gene by nonmetabolized long chain fatty acids. Proc Natl Acad Sci USA 1992;89:10930–10934.
30 Melki SA, Abumrad NA: Glycerolipid synthesis in isolated adipocytes: Substrate dependence and influence of norepinephrine. J Lipid Res 1992;33:669–678.
31 Distel RJ, Robinson GS, Spiegelman BM: Fatty acid regulation of gene expression. Transcriptional and post-transcriptional mechanisms. J Biol Chem 1992;267:5937–5941.
32 Petkovich M: Regulation of gene expression by vitamin A: The role of nuclear retinoic acid receptors. Annu Rev Nutr 1992;12:443–471.
33 Buelt MK, Xu Z, Banaszak LJ, et al: Structural and functional characterization of the phosphorylated adipocyte lipid-binding protein (pp15). Biochemistry 1992;31:3493–3499.
34 Issemann I, Green S: Activation of a member of the steroid hormone receptor superfamily by peroxisome proliferators. Nature 1990;347:645–650.
35 Göttlicher M, Widmark E, Li Q, et al: Fatty acids activate a chimera of the clofibric acid-activated receptor and the glucocorticoid receptor. Proc Natl Acad Sci USA 1992;89: 4653–4657.
36 Dryer C, Krey G, Keller H, et al: Control of the peroxisomal β-oxidation pathway by a novel family of nuclear hormone receptors. Cell 1992;68:879–887.
37 Schmidt A, Endo N, Rutledge SJ, et al: Identification of a new member of the steroid hormone receptor superfamily that is activated by a peroxisome proliferator and fatty acids. Mol Endocrinol 1992;6:1634–1641.
38 Osumi T, Wen JK, Hashimoto T: Two *cis*-acting regulatory sequences in the peroxisome proliferator-responsive enhancer region of rat acyl-CoA oxidase gene. Biochem Biophys Res Commun 1991;175:866–871.
39 Tugwood JD, Issemann I, Anderson RG, et al: The mouse peroxisome proliferator activated receptor recognizes a response element in the 5′ flanking sequence of the rat acyl-CoA oxidase gene. EMBO J 1992;11:433–439.
40 Zhang B, Marcus SL, Sajjadi FG, et al: Identification of a peroxisome proliferator-responsive

element upstream of the gene encoding rat peroxisomal enoyl-CoA hydratase/3-hydroxyacyl-CoA dehydrogenase. Proc Natl Acad Sci USA 1992;89:7541–7545.
41 Muerhoff AS, Griffin KJ, Johnson EF: The peroxisome proliferator-activated receptor mediates the induction of CYP4A6, a cytochrome P-450 fatty acid ω-hydroxylase, by clofibric acid. J Biol Chem 1992;267:19051–19053.
42 Issemann I, Prince R, Tugwood J, et al: A role for fatty acids and liver fatty acid binding protein in peroxisome proliferation. Biochem Soc Trans 1992;20:824–827.
43 Keller H, Dreyer C, Medin J, et al: Fatty acids and retinoids control lipid metabolism through activation of peroxisome proliferator-activated receptor-retinoid X receptor heterodimers. Proc Natl Acad Sci USA 1993;90:2160–2164.
44 Kliewer SA, Umesono K, Nooman DJ, et al: Convergence of 9-*cis* retinoic acid and peroxisome proliferator signalling pathways through heterodimer formation of their receptors. Nature 1992;358:771–774.
45 Green S: Promiscuous liaisons. Nature 1993;361:590–591.
46 Gearing KL, Göttlicher M, Teboul M, et al: Interaction of the peroxisome-proliferator-activated receptor and retinoid X receptor. Proc Natl Acad Sci USA 1993;90:1440–1444.

Gérard P. Ailhaud, Centre de Biochimie (UMR 134 CNRS), Université de Nice-Sophia Antipolis, Faculté des Sciences, Parc Valrose, F-06108 Nice Cedex 2 (France)

Galli C, Simopoulos AP, Tremoli E (eds): Fatty Acids and Lipids: Biological Aspects.
World Rev Nutr Diet. Basel, Karger, 1994, vol 75, pp 46–51

Health Policy Aspects of Lipid Nutrition and Early Development

Norman Salem, Jr., Robert J. Pawlosky

Laboratory of Membrane Biochemistry and Biophysics, Division of Intramural Clinical and Biological Research, National Institutes on Alcohol Abuse and Alcoholism, National Institutes of Health, Rockville, Md., USA

Since the classic work of Burr and Burr [1], it has been accepted that linoleic acid (LA,18:2ω6) is an essential nutrient for the support of growth and development. It is somewhat less clear whether arachidonic acid (AA,20:4ω6) itself is also essential. It has been shown that it has potent 'essential fatty acid (EFA) activity', i.e., that it is potent in substituting for 18:2ω6 where such biological endpoints as growth and transdermal water loss are measured [2, 3]. Hansen et al. [3] have shown that this activity is in part related to retroconversion of 20:4ω6 to 18:2ω6. More recently, Carlson et al. [4] have suggested that compromising 20:4ω6 status may be related to decreases in growth and cognitive scores of premature infants. Linolenic acid (LNA,18:3ω3) also was shown to be effective in restoring growth in EFA-deficient rats but to have far less potency than 18:2ω6 with respect to relieving dermal symptomatology [5]. As Sanders et al. [6] have suggested, to the extent that 18:3ω3 is essential, it represents a requirement for docosahexaenoic acid (DHA,22:6ω3).

Mammalian requirements for dietary 22:6ω3 have of late received a lot of attention and several reviews have been written [7, 8] including several that have focused upon early neural development [2, 9–11]. It has been suggested that adequate sources of 22:6ω3 are needed for proper neural development based largely upon behavioral, electrophysiological and neurological deficits that are associated with inadequate dietary 22:6ω3 or its precursors [11]. Unfortunately, a large-scale dietary experiment has taken place in the Western world in which infants are given formulas devoid of 20- and 22-carbon polyunsaturates such as 22:6ω3. In many cases, the formulas may contain little 18:3ω3 and a high 18:2ω6 to 18:3ω3 ratio, thus exacerbating the problem of

22:6ω3 synthesis and accretion. Salem and co-workers [7, 8, 11–13] have reviewed possible underlying mechanisms involved in 22:6ω3 function.

This discussion then leads to the question of the relative importance of nutritional factors and EFA metabolic capacity in contributing to adequate levels of long chain polyunsaturates (LC PUFAs). Two aspects of this problem will be considered here; these are (1) the use of vegetable oils containing high 18:2ω6 and low 18:3ω3 as the sole fat source during gestation and early development, and (2) the influence of chronic alcohol exposure.

Methods

Domestic female cats (n = 2) of 1.5–2 years of age were fed a diet containing 10 wt% fat composed of 9 wt% corn oil and 1 wt% hydrogenated coconut oil. A second group (n = 2) were fed the same diet except that 20:4ω6 and 22:6ω3 were added so that they accounted for 0.7% each of the total fatty acids. Food and water was available on an ad libitum basis. These diets were given for 4–6 weeks prior to conception and then maintained during pregnancy and lactation. Kittens were weaned to the same diets and sacrificed at 8 weeks of age for brain and retinal analysis.

In a second cat experiment (n = 7), adult males of 4–6 kg weight were fed a diet in which the fat was composed of 1 wt% corn oil and 9 wt% olive oil for 8 months. Half of the animals were given six 1-gram capsules of alcohol once per day for the last 6 months of the dietary experiment. The animals were then sacrificed and the brains and retinas excised for fatty acid analysis.

Rod outer segment membrane preparations were prepared from the kitten retinas according to the method of Stone et al. [14]. Lipids were extracted using the method of Bligh and Dyer [15], methylated with BF_3 in methanol-hexane according to a modification of the method of Morrison and Smith [16] and analyzed by gas chromatography as previously described [17].

Results and Discussion

Vegetable oils such as corn oil frequently serve as the primary dietary fat source in the United States and many other countries in both animal and human diets. An experiment was performed to determine whether a diet rich in this oil could support the 22:6ω3 level in the nervous system of developing kittens. A diet containing 9 wt% corn oil and 1 wt% hydrogenated coconut oil was fed to female cats prior to conception and maintained during pregnancy and lactation. The kittens were weaned to the same diet 6–7 weeks after birth and sacrificed at 8 weeks of age. For a reference point, the same diet supplemented with 20:4ω6 and 22:6ω3 was used. Fatty acid analysis of the diets indicated that the corn oil diet contained 71% of 18:2ω6 and 1.2% of 18:3ω3 and no 20- or 22-carbon PUFAs. The supplemented diet contained the same

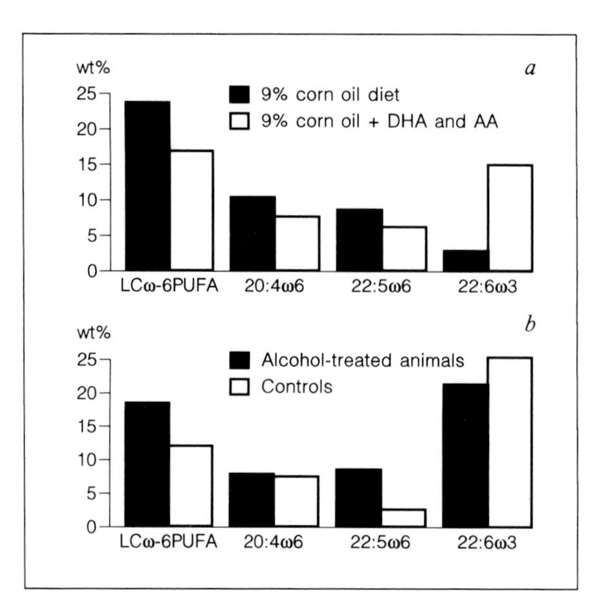

Fig. 1. a Polyunsaturate analysis of 8-week-old kitten retinal rod outer segment preparations after corn oil or corn oil plus AA/DHA diets during gestation and early development. *b* Polyunsaturate analysis of adult cat retinas after 6 months of alcohol.

levels of 18:2ω6 and 18:3ω3 but also contained 0.7% of 20:4ω6, 0.7% of 22:6ω3 and 0.4% of 20:5ω3. This level of LC-PUFA is comparable or somewhat higher than that found in a common cat chow diet.

There was a drastic decline in nervous system 22:6ω3 and a reciprocal increase [18] in long chain ω6 PUFAs such as 20:4ω6, 22:4ω6 and 22:5ω6. In the retinal rod outer segment preparations, this substitution was particularly evident (fig. 1a). Associated with this loss of retinal and brain 22:6ω3 was a delay in the a- and b-wave latencies in the 7-week-old kittens (data not shown). It thus appeared in this pilot experiment that a loss of neural 22:6ω3 was caused by a corn oil based diet and that it was associated with a loss in neural function.

In a second cat experiment, the synergistic effect of a 'barely adequate diet' (BAD) combined with chronic alcohol exposure was used as a model of human alcoholism. Adult male cats were given a single oral dose of ethanol once daily resulting in a peak blood alcohol concentration of 100–120 mg% after about 2 h. Six months of this limited diet and alcohol resulted in a qualitatively similar change in the brain and retina fatty acyl profiles of these animals (fig. 1b). There was a significant decline in 22:6ω3 and a significant increase in 22:5ω6 in the alcohol-exposed animals in comparison to the dietary controls.

Table 1. EFA involvement in fetal alcohol syndrome

Possible causes of LC PUFA level decreases
1. Alcohol lowers LC PUFA formation in the mother
2. Alcohol lowers LC PUFA formation in the fetus
3. Mother's liver functions may be compromised
4. Maternal diet may be inadequate in:
   a. Essential fatty acids
   b. Antioxidant vitamins, minerals and amino acids
5. Alcoholic mother's milk may be deficient in LC PUFA

Possible Effects
1. Nervous system ω3 deficiency syndrome
2. Losses in brain and retinal function
3. Complications in other organs, e.g., the liver

Therapy
1. LC PUFA supplements to pregnant alcoholics may be indicated
2. Clinical and animal research are needed

It thus appeared that alcohol exposure is capable of depleting 22:6ω3 in the adult mammalian nervous system. This is generally not possible as 22:6ω3 is tenaciously retained in the face of dietary insufficiency [2, 7–10, 19] and it is thus rather remarkable. It suggests that alcohol profoundly affects lipid metabolism. Our studies as well as those of others [20, 21] suggest that this is mediated in a complex fashion as it disturbs several metabolic processes related to lipid metabolism. Alcohol is capable of inhibiting EFA elongation and desaturation as well as increasing EFA catabolism. It is unclear which of these mechanisms is of greater quantitative importance in explaining the PUFA-lowering effects of alcoholism.

These studies have important implications for fetal development, particularly for alcoholic mothers. It has been suggested by Holman et al. [22] that pregnancy and lactation are a strain on lipid metabolism so that the mother is not able to maintain her plasma LC PUFA. Our hypothesis is that fetal alcohol exposure results in a 22:6ω3 deficiency that contributes to the neurological damage and loss of functions seen in these infants. There may be a variety of factors contributing to this loss of fetal LC PUFA as summarized in table 1. Decreased EFA elaboration or increased oxidative metabolism in the mother, the placenta and/or the fetus caused by alcohol may lead to lower available pools of LC PUFA and of 22:6ω3 in particular that may be needed for neural development. In addition, alcoholics are believed to have poor diets and the lack of adequate EFA or antioxidant vitamins, minerals or amino acids which serve to maintain LC PUFA status may exacerbate these difficulties. The alcoholic mother may have compromised liver function and this may adversely affect lipid packaging and transport. For a variety of reasons then, it may be predicted that fetal alcohol exposure will compromise the accretion of neural 22:6ω3 and that this will lead to adverse consequences.

Conclusions

Vegetable oil based diets, particularly those with high 18:2ω6 and low 18:3ω3 such as corn, sunflower and safflower oils, do not appear to be adequate fat sources for the support of neural development. They do not support accretion of proper levels of brain and retinal 22:6ω3 and lead to losses in neural performance. It is recommended that maternal diets and artificial milk formulas contain 22:6ω3 and other LC PUFAs. It is rather conservatively suggested that formula levels of LC PUFAs be comparable to those found in human milk in Western women.

In the adult mammal, dietary challenges are known to have little effect on nervous system 22:6ω3 levels. However, chronic alcohol exposure leads to a syndrome that has curious resemblance to the 'ω3 deficiency syndrome' observed during early development when vegetable oils are fed. There is a reciprocal replacement of 22:6ω3 with 22:5ω6. This raises the possibility that the fetal alcohol syndrome has as one of its causes the inadequacy of 22:6ω3 accretion due to the metabolic disturbances caused by alcohol. It also raises the possibility that the cognitive and neurological complications observed in some adult alcoholics may be related to their losses in neural 22:6ω3 and may be treatable with appropriate lipid supplements.

References

1 Burr GO, Burr MM: On the nature of the fatty acids essential in nutrition. J Biol Chem 1930; 86:587–621.

2 Innis SM: Essential fatty acids in growth and development. Prog Lipid Res 1991;30:39–103.

3 Hansen HS, Jensen B, von Wettstein-Knowles P: Apparent in vivo retroconversion of dietary arachidonic to linoleic acid in essential fatty acid-deficient rats. Biochim Biophys Acta 1986; 878:284–287.

4 Carlson SE, Werkman SH, Peeples JM, Cooke RJ, Tolley EA, Wilson WM III: Growth and development of very-low birthweight infants in relation to ω3 and ω6 essential fatty acid status; in Sinclair A, Gibson R (eds): Essential Fatty Acid and Eicosanoids. Third International Conference. Champaign, American Oil Chemists' Society, 1992, pp 192–197.

5 Lundberg WO: On the quantification of essential fatty acid requirements. Fette-Seifen-Anstrichmittel 1979;81:337–348.

6 Sanders TAB, Mistry M, Naismith DJ: The influence of maternal diet rich in linoleic acid on brain and retinal docosahexaenoic acid in the rat. Br J Nutr 1984;51:57–66.

7 Salem N Jr, Kim HY, Yergey J: Docosahexaenoic acid: Membrane function and metabolism; in Simopoulos AP, Kifer RR, Martin R (eds): The Health Effects of Polyunsaturates in Seafoods. New York, Academic Press, 1986, pp 263–317.
8 Salem N Jr: Omega-3 fatty acids: Molecular and biochemical aspects; in Spiller G, Scala J (eds): New Protective Roles of Selected Nutrients. New York, Liss, 1989, pp 109–228.
9 Neuringer M, Connor WE: ω3 Fatty acids in the brain and retina: Evidence for their essentiality. Nutr Rev 1986;44:285–294.
10 Uauy R, Birch E, Birch D, Peirano P: Visual and brain function requirements in studies of ω3 fatty acid requirements of infants. J Pediatr 1992;120(suppl):168–180.
11 Salem N Jr, Ward GR: Are ω3 fatty acids essential nutrients for mammals? World Rev Nutr Diet 1993;72:128–147.
12 Salem N Jr, Niebylski CD: An evaluation of alternative hypotheses involved in the biological function of docosahexaenoic acid in the nervous system; in Sinclair A, Gibson R (eds): Essential Fatty Acids and Eicosanoids. Champaign, American Oil Chemists' Society, 1992, pp 84–86.
13 Salem N Jr, Pawlosky RJ: Docosahexaenoic acid is an essential nutrient in the nervous system; in Kobayashi T (ed): Proc 1st Int Conf Vitamins and Biofactors in Life Science, Kobe. J Nutr Sci Vitaminol. Tokyo, Center for Academic Publications Japan, 1992, pp 153–156.
14 Stone WL, Farnsworth CC, Dratz EA: A reinvestigation of the fatty acid content of bovine, rat and frog retinal rod outer segments. Exp Eye Res 1979;28:387–397.
15 Bligh EG, Dyer WJ: A rapid method of total lipid extraction and purification. Can J Biochem Phys 1959;37:911–917.
16 Morrison WR, Smith LM: Preparation of fatty acid methyl esters and dimethylacetals from lipids with boron fluoride-methanol. J Lipid Res 1964;5:600–608.
17 Knapp HR, Salem N Jr: Formation of PGI_3 in the rat during dietary fish oil supplementation. Prostaglandins 1989;38:509–521.
18 Galli C, Trzeciak HI, Paoletti R: Effects of dietary fatty acids on the fatty acid composition of brain ethanolamine phosphoglyceride: Reciprocal replacement of ω6 and ω3 polyunsaturated fatty acids. Biochim Biophys Acta 1971;248:449–454.
19 Tinoco J: Dietary requirements and functions of alpha-linolenic acid in animals. Prog Lipid Res 1982;21:1–45.
20 Salem N Jr: Alcohol, fatty acids and diet. Alcohol Health Res World 1989;13:211–218.
21 Salem N Jr, Ward GR: The effects of ethanol on polyunsaturated fatty acid composition; in Alling C, Sun GY (eds): Alcohol, Cell Membranes and Signal Transduction in Brain. New York, Plenum Press, 1993.
22 Holman RT, Johnson SB, Ogburn PL: Deficiency of essential fatty acids and membrane fluidity during pregnancy and lactation. Proc Natl Acad Sci USA 1991;88:4835–4839.

Norman Salem, Jr., MD, LMBB, NIAAA, Room 55C, 12501 Washington Avenue, Rockville, MD 20852 (USA)

Galli C, Simopoulos AP, Tremoli E (eds): Fatty Acids and Lipids: Biological Aspects.
World Rev Nutr Diet. Basel, Karger, 1994, vol 75, pp 52–62

Significance of ω3 Fatty Acids for Retinal and Brain Development of Preterm and Term Infants [1]

Ricardo Uauy-Dagach, [a,b,c], *Eileen E. Birch* [b], *David G. Birch* [b], *Dennis R. Hoffman* [b,c]

[a] Clinical Nutrition Unit, Institute of Nutrition and Food Technology (INTA), University of Chile, Santiago, Chile;
[b] Retina Foundation of the Southwest, Dallas, Tex;
[c] Division of Neonatology, Department of Pediatrics, University of Texas Southwestern Medical Center at Dallas, Tex., USA

Until recently, pediatricians considered essential fatty acids (EFAs) mainly as part of the lipid energy supply necessary for growth, cellular metabolism, and muscle activity. Further significance was gained with the knowledge that EFAs served as dietary precursors for eicosanoid and docosanoid formation. During this decade, attention has focused on the role of ω3 EFAs in the prevention of cardiovascular disease, immune disorders and in their potential key role as structural components of membrane phospholipids necessary for normal eye and brain development [1–5].

Over 30 years ago, Hansen et al. [6] firmly established linoleic acid (LA (18:2 ω6)) as essential for normal infant nutrition in a clinical and biochemical study of 428 infants fed cow's milk-based formulations with different types of fat. Daily LA intake of study infants ranged from 10 mg/kg while fed a fully skimmed milk preparation to 800 mg/kg when a corn/coconut oil-based preparation was fed. They observed dryness, desquamation and thickening of the skin and growth faltering as the most frequent manifestations of LA deficiency in young infants.

[1] This work was supported by NIH Grants HD 22380, EY 05235 and EY 05236 and by Fondecyt Chile Grant 1930820.

Our interest over the past 7 years has focused on the potential requirements and benefits of the dietary ω3 EFAs, α-linolenic acid (LNA (18:3 ω3)) and docosahexaenoic acid (DHA (22:6 ω3)), for optimal infant visual and brain development [7–11]. These studies were designed to investigate in human infants the questions posed by the observations of Neuringer and Connor [12, 13] using primate models of ω3 fatty acid (FA) deficiency. These investigators developed a pre- and postnatal ω3 deficiency model using safflower oil in which the amount of LNA is very low and the ratio of LA to LNA is very high (approximately 250:1). Similarly, we used preterm infants fed a commercial corn oil-based formula as a model for ω3 deficiency. At the time of our initial studies, 1986, all infants received ample LA but virtually no LNA, the ratio of LA to LNA in most infant formulas was close to 50:1.

The purpose of our studies was to assess the essentiality of ω3 FA in humans and specifically for very-low-birth-weight (VLBW) infants who we suspected were particularly vulnerable to deficiency given the virtual absence of adipose tissue stores at birth, the possible immaturity of the FA elongation/desaturation pathways and the inadequate LNA and DHA intake provided by formula. Over the past decade, we and others have conducted studies to evaluate the effect of ω3 FA supplementation in VLBW infants examining the effects of LNA or LNA plus DHA supplementation on plasma and tissue lipid composition, retinal electrophysiologic function, the maturation of the visual cortex and measures of infant growth and development [7–11, 14–16].

In a series of studies, we have characterized the biochemical and functional effects of dietary ω3 FA deficiency in preterm infants. When we initiated our studies, low LNA was found in powdered infant formulas in most countries; however, by now virtually all infant formulas are supplemented with LNA and some manufacturers in Europe have added DHA to preterm formulas. The European Society for Pediatric Gastroenterology and Nutrition has recommended not only that LNA be present but has stated that it would be desirable that DHA and arachidonic acid (AA) also be added in formulas destined for preterm infants [17]. Despite efforts by many investigators, the American Academy of Pediatrics/USA still has not acknowledged the need for and the essentiality of ω3 FAs; even today low LNA formulas are still in use in some parts of the world [18]. In addition, we and others have demonstrated that long-chain polyunsaturated fatty acids (LCPUFAs), such as DHA, provide a specific structural environment within the phospholipid bilayer; thus, influencing important membrane functions as ion or solute transport, receptor activity or enzyme action [19–22].

We chose to assess visual function since it has been shown to serve as a sensitive index of the subtle actions of ω3 FAs on visual development of vertebrates. Animal studies clearly indicate that the high DHA content of neural tissues such as retina and cerebral cortex is of functional importance, and that replacement of this lipid by FAs of the ω6 or ω9 families impairs visual and brain function [12, 13, 23, 24]. A significant correlation between brain and red blood cell (RBC) membrane phospholipid DHA composition during deficiency was also established in animal studies, thus permitting an evaluation of dietary intervention on DHA status in the central nervous system from analysis of blood phospholipids [13, 25, 26].

We studied 83 newborns with body weights of 1,000–1,500 g, 28–32 weeks' gestation at birth, receiving enteral feedings and free of major neonatal morbidity by day 10 of life. Ten infants receiving human milk (HM) served as controls for the study; we also evaluated 12 infants who remained in utero until 35 weeks' conceptional age and compared them to the study infants. The HM-fed group was supplemented with HM fortifier (Enfamil Mead-Johnson) and received ⩾75% of their intake as own mother's HM up to 36 weeks' postconceptional age (usual discharge age); if mothers were unable to fully provide HM, the soy/marine oil formula was used to supplement their feeding. The remaining 73 infants were randomly assigned to three formula groups varying in EFA. Experimental formula feedings began at day 10 of life and continued until 57 weeks' postconceptional age (i.e., equivalent to 4 months postterm). Full details on study subjects and experimental protocol can be found elsewhere [8].

The EFA composition of HM and study formulas has been published in detail. Briefly the corn oil-based formula corresponding to commercial powdered premature formula and contained 24% LA, was extremely low in LNA (0.4%) and had no LCPUFA. The soy oil-based formula was equivalent to present ready-to-feed premature formula and supplied 21% LA and 2.7% LNA as precursors of LCPUFAs. The third experimental formula was also soy oil-based, 1.4% LNA, but was enriched with marine oil to give an ώ3 LCPUFA (0.6% EPA and 0.4% DHA) content closer to that found in HM [8]. At the time we started our study, ώ3 LCPUFA supplementation could only be obtained from marine oils. We chose to include sufficient oil to mimic the DHA content of HM since we were uncertain on how effectively preterm infants could convert EPA to DHA. The full nutritional content of the preterm milk formula has been published [7, 8].

Full-field electroretinographic (ERG) responses to short and long wavelength stimuli over an extensive range of retinal illuminances were evaluated at 36 and 57 weeks' postconception. The ERG response to short wave stimuli was used to help isolate rod function, at the two test ages. Naka-Rushton plots were computed from rod responses to graded illuminances; threshold, maximum amplitude and semisaturation constants were calculated. The use of long and short wavelength stimuli in dark-adapted infants to separate retinal cone and rod ERG permitted the evaluation of the differential diet effect on the function of cones and rods. These methods have been described elsewhere and complete results of our ERG function studies have been recently published [7, 9, 27–29].

Visual acuity development measured by pattern-reversal visual-evoked potentials (VEP) was evaluated at the 36- and 57-week follow-up. Behavioral measurements of visual acuity, forced-choice preferential looking (FPL) visual acuity responses, could be reliably tested only at the 57-week follow-up. Since the VEP and FPL methods differ, the acuity values obtained using these methods are not identical. Details of the methodology and results have been recently published [10, 28, 30]. The VEP and FPL visual acuity tests can be used to evaluate the neural integrity of the pathway from the retina to the primary visual cortex pathway and the ability of the infant to see and produce a motor response (fixation) when visually stimulated [30, 31].

Compositions of RBC lipids for all diet groups were similar on entry into the study [8]. At the 36- and 57-week postconception follow-up, marked differences in DHA content were evident in the RBC lipids of all diet groups. Infants fed the ω3 FA-deficient corn oil formula had significant reductions in DHA compared to the other formula groups. The marine oil group presented DHA concentrations elevated above all other groups. The ω6 docosapentaenoic acid (DPA) relative content was found at 36 and 57 weeks in RBC lipids to be inversely related to the DHA content. DPA was significantly elevated in the corn oil group in comparison to the HM-fed and soy/marine oil groups. By 57 weeks, the differences in the content of the end products of ω3 and ω6 FA metabolism, namely DHA and DPA, were greatly accentuated compared to 36-week values. The corn oil-fed group was significantly different in both DHA and DPA content compared to all other diet groups. RBC lipids of the HM and marine oil groups were statistically different in DHA composition but showed a similar pattern across total ω3 and ω6 LCPUFA. The soy oil group had intermediate DHA and DPA values relative to corn and soy/marine oil groups but differed significantly from the DHA-supplemented group [8, 32].

Retinal function responses demonstrated significantly higher threshold values in rod photoreceptors from the ω3-deficient group (corn oil-fed) at the 36 week follow-up relative to groups receiving ω3 FA [7, 8]. Figure 1 depicts graphically the correlation between plasma DHA relative content and rod threshold; that is, the light intensity required to elicit a 2-μV response; a significant negative correlation indicates that more light was required when plasma DHA levels were low. Figure 2 shows the positive correlation between plasma DHA levels and the maximal amplitude in the b-wave response to light. The higher threshold and lower maximum amplitude values when the DHA content is low indicates that the sensitivity and maturity of the rods of the ω3 FA-deficient infants are significantly reduced when compared to the infants fed soy/marine formula or HM. A group of 10 infants, born at 35 weeks, had their

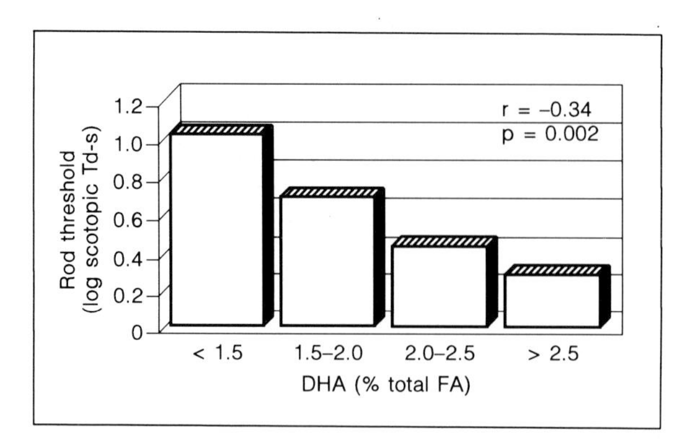

Fig. 1. Correlation between DHA content of plasma and rod threshold derived from Naka-Rushton plots at 36 weeks after conception. Threshold is the light intensity measured in log scotopic Td-s (troland seconds) required to induce a b-wave amplitude $> 2\ \mu V$. Nonlinear regression r value and corresponding p are included. Less light was required when DHA content was high. [For detailed results, see 9.]

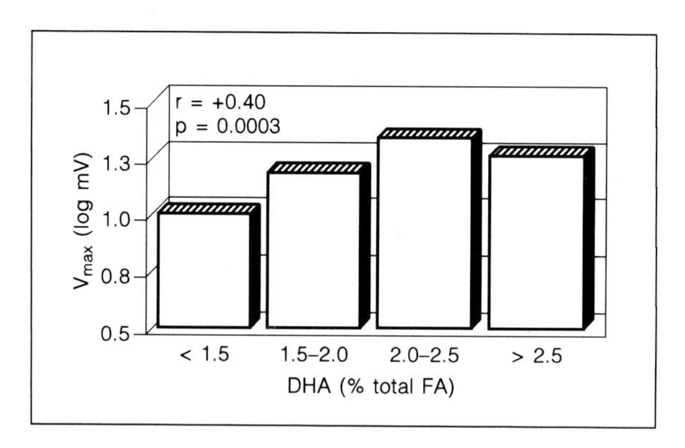

Fig. 2. Correlation between DHA content of plasma and rod maximum amplitude derived from Naka-Rushton plots at 36 weeks after conception. V_{max} is the maximal b-wave amplitude measured in microvolts achieved at the highest light intensity tested. Nonlinear regression r value and corresponding p are included. Higher V_{max} values were obtained when DHA content was high. [For detailed results, see 9.]

visual maturity tested 3–5 days after birth and served as a normal standard for comparison. This 'in utero' group of infants receiving transplacentally acquired EFAs were nearly identical in all rod ERG functional indices to the equivalent conceptional age HM and marine oil-fed premature infants. The soy oil-fed group had marginally higher threshold values than the soy/marine oil-fed group (p value was <0.06). Cone function was not significantly affected by diet, although the trends were similar to the results obtained from rod photorecep-

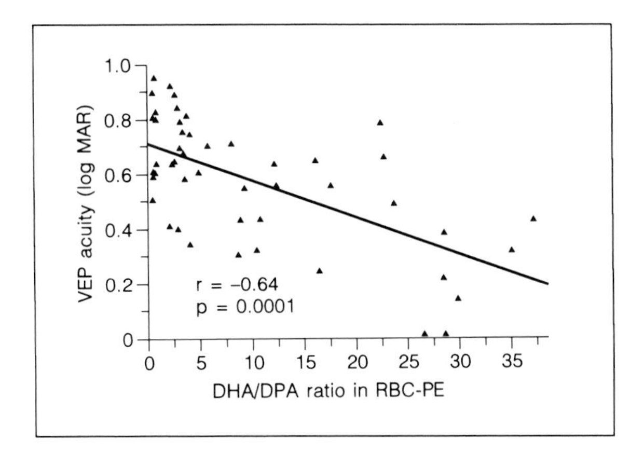

Fig. 3. Correlation between DHA/DPA ratio in red cells and visual acuity measured by pattern reversal VEP expressed as log MAR (minimal angle of resolution). A log MAR value of 0 corresponds to 20/20 Snellen equivalents while a value of 1 corresponds to 20/200. A significantly negative linear correlation was found between DHA/DPA in phosphatidylethanolamine of RBCs and log MAR values (low log MAR indicates better acuity). High DHA levels were associated to better acuity. [For detailed results, see 10.]

tors. At the 57-week follow-up, retinal rods and cones showed no diet-induced differences in a- or b-wave parameters. The corn oil-fed group at 57 weeks consistently had longer implicit times in light-adapted oscillatory potentials. Some peaks of the oscillatory potentials of the soy oil group were significantly different from the HM-fed infants while the soy/marine never differed [9]. Oscillatory potentials are generated in the inner retina and probably reflect ganglion cell-amacrine cell interactions.

The soy/marine and the HM-fed groups had lower values of log MAR (minimal angle of resolution), that is better acuity, relative to infants fed formulas devoid of DHA at the 57-week follow-up using either VEP (electrophysiologic response) or the FPL (behavioral) methods. A log MAR value of 0 corresponds to 20/20 Snellen equivalents while a value of 1 corresponds to 20/200. A significantly negative correlation was found between DHA levels and log MAR values (low log MAR indicate better acuity), that is, when DHA levels were high, log MAR values were low. Figure 3 and 4 demonstrate graphically the correlation between VEP acuity and red cell DHA/DPA ratio in two phospholipid fractions. The HM and soy/marine oil-fed infants had the highest DHA/DPA ratio and also had the best visual acuity (lowest log MAR). The group that received soy oil as a source of ω3 FA was different in VEP acuity at the 57-week follow-up relative to the soy/marine oil-fed group, indicating that despite ample LNA, function was not the same as in the DHA-supplemented

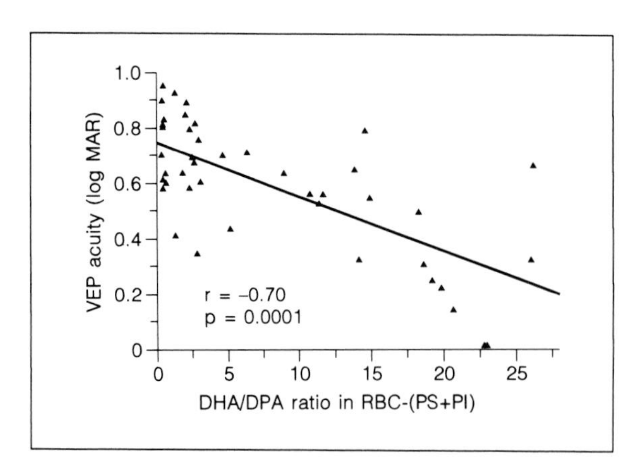

Fig. 4. Correlation between DHA/DPA ratio in red cells and visual acuity measured by pattern reversal VEP expressed as log MAR (minimal angle of resolution). A log MAR value of 0 corresponds to 20/20 Snellen equivalents while a value of 1 corresponds to 20/200. A significantly negative linear correlation was found between DHA/DPA of phosphatidylserine+ phosphatidylinositol in RBCs and log MAR values (low log MAR indicate better acuity). High DHA levels were associated to better acuity. [For detailed results, see 10.]

group. A group of healthy full-term breast-fed infants matched by conceptional age were used as controls. The HM-fed and soy/marine oil-fed preterm groups were virtually identical to these controls while the corn and soy oil groups had significantly poorer visual acuities [10, 28].

Most observations in term infants are limited to studies on the effect of diet on blood lipid profiles. the concentration of DHA in the RBCs of full-term infants fed formula has been found to be consistently lower than that of breast-fed infants [4, 33]. Infants fed LNA-supplemented formula do not reach the DHA concentrations of breast-fed infants. This indicates that present commercial formulas provide insufficient LNA or that chain-elongation/desaturation enzymes are not sufficiently active during initial postnatal life to support tissue accretion of DHA. The functional significance of this biochemical change for full-term infants is beginning to be unraveled.

We have conducted preliminary studies of visual function [10] in full-term infants fed either HM ($n = 18$) or cow's milk formula ($n = 12$) with 12–18% LA and 0.5–1.0% LNA. We found that both VEP and FPL acuities were higher in 4-month-old exclusively breast-fed infants compared to formula-fed infants. The visual acuities, expressed in Snellen equivalents, for VEP were significantly worse for the formula-fed infants (mean value 20/83) compared to the HM-fed group (20/65, $p < 0.05$). Similarly, FPL acuities were worse in the formula-fed group (20/129) than in the HM-fed infants (20/107, $p < 0.025$) indicating improved vision in the term infants receiving dietary LCPUFAs.

We have also completed a 3-year follow-up of 43 healthy full-term infants recruited from a Dallas high socioeconomic status private practice into a study of controlled lipid intake for the entire first year of life [11]. The two cohorts of full-term infants were fed from birth; exclusively breast milk for at least 4 months, or commercial formula from birth through 12 months (LNA $<0.6\%$ of total energy). The breast-fed group was weaned to an oleate predominant formula and received egg yolk for the remainder of the first year. The HM-fed infants maintained higher DHA concentrations in their plasma and RBC membrane phospholipids throughout their first year relative to the formula group. When tested for visual function at 3 years of age, differences were noted between the breast-fed and the formula groups in stereoacuity measured using operant preferential looking (OPL) methodology. Mean OPL $\pm$ SD stereoacuity was 42 ± 5 s for the breast-fed group while it was over twice that value for the formula-fed infants. Since values were not normally distributed, nonparametric testing was done showing that 92% of the breast-fed group had fully mature OPL stereoacuity (40 s or better) whereas only 35% of the formula-fed infants met that criterion ($p<0.001$). Monocular acuities assessed by OPL were similar in both groups; mean Snellen equivalents were 20/28 for the HM- and 20/32 for the formula-fed infants ($p<0.16$). Visual recognition was tested by the child's ability to match 3 letters. The mean matching score was 2.71 for the breast-fed and 1.82 for the formula-fed infants ($p<0.04$). The proportion of children that matched 3 of 3 letters presented was 90% for the breast-fed while it was only 61% for the formula-fed group ($p<0.001$). No differences were found in picture-naming or in color vision. The potential for greater consequences of these sensory differences in less privileged settings is easily perceived from these results. In a setting of psychosocial deprivation, minor differences in sensory and cognitive functions are usually amplified. A recent report from Australia provides further evidence supporting our initial finding that HM-fed term infants have improved visual acuity relative to formula-fed infants despite the presence of 1–1.6% of LNA in the fat of the formula [34]. Furthermore, the investigators report that in the 16 infants studied at 5 months of age, there was a significant correlation between DHA content in RBCs and visual acuity, measured using pattern-reversal VEP.

In 1929, Burr and Burr [25] introduced the concept that specific components of fat may be necessary for the proper growth and development of animals and possibly humans. EFAs were considered of marginal nutritional importance until the 1960s when signs of clinical deficiency became apparent in infants fed skim milk-based formulas and in those given lipid-free parenteral nutrition [6, 36–38]. During this decade it is becoming generally accepted that ω3 FAs play a key role in normal infant brain and eye development.

The biochemical data on lipid composition of plasma and RBC membrane lipids document that indeed DHA-containing diets produce discernible changes in the ω3 FA content of blood and tissue components. This serves not only to demonstrate compliance but, since measures of visual function and FA composition are correlated to the indices of eye and brain function, it provides powerful evidence suggesting that it is the ω3 FA supply of the early diet which is responsible for enhanced visual maturation. The reversibility of changes in visual function cannot be fully answered from our data since the study was terminated at 57 weeks' postconception. Most of the ERG changes at 36 weeks were, in fact, reversible based on the 57-week results. However, this cannot be answered for the visual acuity data since no long-term follow-up data from randomized trials are presently available.

The possibility for long-term effects cannot be discarded easily since a study in infants born at term, dying from sudden infant death, revealed that brain composition is affected by HM feeding in terms of higher DHA and lower DPA content relative to infants receiving cow's milk-based formulas [39]. The lowest DHA content in brain cortex phospholipids was observed in those fed formulas with a high LA to LNA ratio. Further evidence in support of long-term consequences of early HM is provided by results from a randomized controlled trial of preterm infants indicating that HM feeding by nasogastric tube for 30 days was associated with a +8.3 point IQ difference at 8 years of age relative to a formula-fed group, after controlling for socioeconomic and other maternal variables [40]. These are the first controlled observations indicating that HM may offer unique advantages for brain development.

The information from full-term infants is limited by the impossibility of conducting true controlled trials of HM feeding. The results from several studies of large cohorts comparing the effect of HM feeding indicate that infants receiving formula have poorer pattern recognition, lower scores in pictorial, verbal and mathematical attainment tests, and lower overall intelligence quotients [11, 41]. Whether ω3 FAs play a role in these phenomena remains to be determined. Presently several controlled trials of LCPUFA supplementation of formula-fed term infants are underway. The answer should be forthcoming.

The safety aspects of our studies included measures of growth, bleeding function, RBC membrane properties, vitamin A and E levels, and measures of lipid peroxidation [7, 42, 43]. No significant adverse diet-induced effects were evident in any of these indices and all values were within the acceptable ranges by clinical standards. The limited sample size detracts from using these data as conclusive evidence for the safety of fish oil supplementation. Alternate sources of ω3 FAs are presently being investigated in several centers. Larger scale clinical trials are in order to fully satisfy safety concerns.

Our studies provide clear evidence that dietary ω3 FA deficiency will affect eye and brain function of preterm infants as measured by ERGs, cortical VEP and behavioral testing of visual acuity. Preterm infants require DHA in their diet because they are unable to form it in sufficient quantity from LNA provided by soy oil-based formula products. Changes in ω3 and ω6 FA intake resulted in discernible differences in the FA composition of plasma lipid and RBC membranes. We conclude that the supply of EFAs provided by the early diet influences the maturation of visual and brain function. Our studies summarized in this article support an essential role for ω3 LCPUFA for optimal visual development in preterm and term infants.

References

1 Sprecher H: Biochemistry of essential fatty acids. Prog Lipid Res 1981;20:13–22.
2 Bazan NG: The metabolism of omega-3 polyunsaturated fatty acids in the eye: The possible role of docosahexaenoic acid and docosanoids in retinal physiology and ocular pathology. Prog Clin Biol Res 1989;312:95–112.
3 Willis AL: Essential fatty acids, prostaglandins, and related eicosanoids; in Olson RE (ed): Present Knowledge in Nutrition. Washington, The Nutrition Foundation Inc, 1984, pp 90–113.
4 Uauy R, Hoffman DR: Essential fatty acid requirements for normal eye and brain development. Semin Perinatol 1991;15:449–455.
5 Simopoulos AP: Omega-3 fatty acids in health and disease and in growth and development. Am J Clin Nutr 1991;54:438–463.
6 Hansen AE, Wiese HF, Boelsche AN, et al: Role of linoleic acid in infant nutrition: Clinical and chemical study of 428 infants fed on milk mixtures varying in kind and amount of fat. Pediatrics 1963;31:171–192.
7 Uauy RD, Birch DG, Birch EE, et al: Effect of dietary omega-3 fatty acids on retinal function of very-low-birth-weight neonates. Pediatr Res 1990;28;485–492.
8 Hoffman D, Uauy R: Essentiality of dietary omega-3 fatty acids for premature infants: Plasma and red blood cell fatty acid composition. Lipids 1992;27:886–895.
9 Birch DG, Birch EE, Hoffman DR, et al: Retinal development in very-low-birth-weight infants fed diets differing in omega-3 fatty acids. Invest Ophthalmol Vis Sci 1992;33:2365–2376.
10 Birch EE, Birch DG, Hoffman DR, et al: Dietary essential fatty acid supply and visual acuity development. Invest Ophthal Vis Sci 1992;33:3242–3253.
11 Birch E, Birch D, Hoffman D, et al: Breast-feeding and optimal visual development. J Pediatr Ophthalmol Strabismus 1993;30:33–38.
12 Neuringer M, Connor WE, Van Petten C, et al: Dietary omega-3 fatty acid deficiency and visual loss in infant rhesus monkeys. J Clin Invest 1984;73:272–276.
13 Neuringer M, Connor WE, Lin DS, et al: Biochemical and functional effects of prenatal and postnatal omega-3 fatty acid deficiency on retina and brain in rhesus monkeys. Proc Natl Acad Sci USA 1986;83:4022–4025.
14 Koletzko B, Schmidt E, Bremer HJ, et al: Effects of dietary long-chain polyunsaturated fatty acids on the essential fatty acid status of premature infants. Eur J Pediatr 1989;148:669–675.
15 Innis SM, Foote KD, MacKinnon MJ, et al: Plasma and red blood cell fatty acids of low-birth-weight infants fed their mother's expressed breast milk or preterm infant formula. Am J Clin Nutr 1990;51:994–1000.
16 Carlson SE, Cooke RS, Rhodes PG, et al: Effect of vegetable and marine oils in preterm infant formulas on blood arachidonic and docosahexaenoic acids. J Pediatr 1991;120:S159–S167.
17 European Society of Paediatric Gastroenterology and Nutrition Committee on Nutrition: Comment on the content and composition of lipids in infant formulas. Acta Paediatr Scand 1991;80: 887–896.

18 Hansen J: Appendix: Commercial formulas for preterm infants 1992; in Tsang RC, Lucas A, Uauy R, Zlotkin S (eds): Nutritional Needs of Preterm Infants: Scientific Basis and Practical Guidelines. New York, Williams & Wilkins Caduceus Medical Publishers Inc, 1993, pp 297–301.
19 Stubbs CD, Smith AD: The modification of mammalian polyunsaturated fatty acid composition in relation to fluidity and function. Biochim Biophys Acta 1984;779:89–137.
20 Dratz E, Deese A: The role of docosahexaenoic acid (22:6 ω3) in biological membranes; in Simopoulos AP, Kiffer RR, Martin RE (eds): Health Effects of Polyunsaturated Fatty acids in Seafoods. New York, Academic Press, 1986, pp 319–351.
21 Salem N Jr, Kim HY, Yergey JA: Docosahexaenoic acid: Membrane function and metabolism; in Simopoulos AP, Kiffer RR, Martin RE (eds): Health Effects of Polyunsaturated Fatty Acids in Seafoods. New York, Academic Press, 1986, pp 263–317.
22 Treen M, Uauy RD, Jameson DM, et al: Effect of docosahexaenoic acid on membrane fluidity and function in intact cultured Y-79 retinoblastoma cells. Arch Biochem Biophys 1992;294:564–570.
23 Wheeler TG, Benolken RM, Anderson RE: Visual membranes: Specificity of fatty acid precursors for the electrical response to illumination. Science 1975;188:1312–1314.
24 Bourre JM, Francois M, Youyou A, et al: The effects of dietary α-linolenic acid on the composition of nerve membranes, enzymatic activity, amplitude of electrophysiological parameters, resistance to poisons and performance of learning tasks in rats. J Nutr 1989;119:1880–1892.
25 Carlson SE, Carver JD, House SG: High fat diets varying in ratios of polyunsaturated to saturated fatty acid and linoleic to linolenic acid: A comparison of rat neural and red cell membrane phospholipids. J Nutr 1986;116:718–726.
26 Connor WE, Lin DS, Neuringer M: Is docosahexaenoic acid (DHA, 22:6 ω3) content of erythrocytes a marker for the DHA content of brain phospholipids? FASEB J 1993;7:152A.
27 Birch EE, Birch DG, Petrig B, et al: Retinal and cortical function of infants at 36 and 57 weeks postconception. Clin Vision Sci 1990;5:363–373.
28 Uauy R, Birch E, Birch D, et al: Visual and brain function measurements in studies on ω3 fatty acid requirements of infants. J Pediatr 1992;120:S168–S180.
29 Birch DG: Clinical electroretinography. Ophthalmol Clin North Am 1989;2:469–497.
30 Birch EE: Visual acuity testing in infants and young children. Ophthalmol Clin North Am 1989;2: 369–389.
31 Sokol S: Visually evoked potentials: Theory, techniques, and clinical applications. Surv Ophthalmol 1976;21:18–44.
32 Hoffman DR, Birch EE, Birch DG, et al: Effects of ω3 long-chain polyunsaturated fatty acid supplementation on retinal and cortical development in premature infants. Am J Clin Nut 1993; 57:807S–812S.
33 Uauy R, Treen M, Hoffman D: Essential fatty acid metabolism and requirements during development. Semin Perinatol 1989;13:118–130.
34 Makrides M, Simmer K, Goggin M, et al: Erythrocyte docosahexaenoic acid correlates with the visual response of healthy, term infants. Pediatr Res 1993;34:425–427.
35 Burr GO, Burr MM: A new deficiency disease produced by rigid exclusion of fat from the diet. J Biol Chem 1929;82:345–367.
36 Caldwell MD, Johnsson HT, Othersen HB: Essential fatty acid deficiency in an infant receiving prolonged parenteral alimentation. J Pediatr 1972;81:894–898.
37 Paulsrud JR, Pensler L, Whitten CF, et al: Essential fatty acid deficiency in infants induced by fat-free intravenous feeding. Am J Clin Nutr 1972;25:897–904.
38 White HB, Turner MD, Turner AC, et al: Blood lipid alterations in infants receiving intravenous fat-free alimentation. J Pediatr 1973;83:305–313.
39 Farquharson J, Cockburn F, Ainslie PW, et al: Infant cerebral cortex phospholipid fatty acid composition and diet. Lancet 1992;340:810–813.
40 Lucas A, Morley R, Cole TJ, et al: Breast milk and subsequent intelligence quotient in children born preterm. Lancet 1992;339:261–264.
41 Rodgers B: Feeding in infancy and later ability and attainment: A longitudinal study. Dev Med Child Neurol 1978;20:421–426.
42 Treen M, Hoffman D, Jameson D, et al: Effect of dietary essential fatty acids on membrane properties of intact RBCs in VLBW neonates. Pediatr Res 1989;25:298A.
43 Buchanan GR, Uauy R, Holtkamp C, et al: Bleeding time measurements in healthy very low birth weight neonates between 10 and 120 days of age. Pediatr Res 1990;27:262A.

Ricardo Uauy-Dagach, MD, PhD, INTA University of Chile, Casilla 138–11, Santiago (Chile)

Galli C, Simopoulos AP, Tremoli E (eds): Fatty Acids and Lipids: Biological Aspects.
World Rev Nutr Diet. Basel, Karger, 1994, vol 75, pp 63–69

Growth and Development of Premature Infants in Relation to ω3 and ω6 Fatty Acid Status

Susan E. Carlson[a, b], *Susan H. Werkman*[a], *Jeanette M. Peeples*[a], *William M. Wilson, III*[c]

Departments of [a]Pediatrics and [b]Obstetrics and Gynecology, and [c]Boling Center for Developmental Disabilities, The University of Tennessee, Memphis, Tenn., USA

The ω6 fatty acid, arachidonic acid (AA), and the ω3 fatty acid, docosahexaenoic acid (DHA), account for approximately 50% of the total fatty acids in the gray matter of the brain [1]. Clandinin et al. [2] have shown that approximately 80% of intrauterine AA and DHA accumulation occurs during the last intrauterine trimester. Martinez et al. [3] have confirmed the rapid last trimester accumulation and shown that this rate of accumulation continues in the months following normal term birth. Infants accumulate AA and DHA from mother by intrauterine transfer and from human milk which contains these long chain metabolites [4, 5]. In contrast, formulas prepared for feeding term and preterm infants in the USA contain only the 18 carbon essential fatty acids of the ω6 and ω3 families, linoleic acid (LA) and linolenic acid (LNA), respectively [4, 5].

Plasma and red blood cell phospholipid AA and DHA remain high during fetal life and during human milk feeding [4, 5] but fall during formula feeding despite good amounts of LA and LNA [4–7]. We have taken this as evidence of declining AA and DHA status in formula-fed compared to human milk-fed infants. The recent report that brain DHA is higher in human milk-fed than formula-fed infants confirms our early belief and provides an important missing link [8]. It is well known that rats and monkeys deprived of normal neural and retinal DHA accumulation during development have abnormal retinal physiology [9, 10]. Rodent behavior is also influenced by ω3 deprivation in rats [11, 12] and visual acuity development in monkeys [13]. Our hypothesis has been that DHA is also a conditionally essential nutrient for the visual and early cognitive development of preterm infants.

Table 1. Study I goals

1. To maintain DHA status in the range of human milk-fed infants
2. To test the hypothesis that this would improve visual acuity
3. To test the hypothesis that this would improve early indices of cognitive development
4. To monitor ω6 fatty acid (AA) status, growth, psychomotor development and vitamin A status

Table 2. Study I conclusions

1. Marine oil (Marine I) maintained DHA in the range of human milk feeding [7]
2. Marine I improved visual acuity at 2 and 4 months past term [14, 15]
3. Visual acuity was predicted by DHA status even within the group of infants fed Marine I [15]
4. Marine I decreased AA status [7] and growth achievement [16]
5. AA status was highly correlated with growth [17]
6. The Bayley MDI was not influenced by diet (table 4)
7. The Bayley PDI tended to be reduced in infants fed Marine I compared to control formula (table 5) and this was related to reductions in linear growth achievement [18]
8. Nutritional status appeared to be marginal in both diet groups due to discharge at 1.8 kg on term formula [19, 20]

During the last 10 years, we have completed a series of studies in preterm infants to determine the effects of a dietary source of DHA ω3 on ω3 and ω6 fatty acid status, visual acuity, growth and development. The results of two studies in which infants were given experimental formulas containing DHA are reported here. In study I, a marine oil source of DHA (Marine I) was fed as 0.2% of total fatty acids from approximately 3 weeks of age through 9 months past expected term. Marine I contained eicosapentaenoic acid (EPA) as 0.3% of total fatty acids. Experimental and control formulas were provided in a nutrient-enriched preterm formula until about 1.8 kg and in a term formula thereafter. The goals and conclusions of this published study are provided as a summary with references to the original articles in tables 1 and 2.

In study II, which was completed in October 1993, the experimental formula contained 0.2% DHA and 0.03% EPA (Marine II) and was fed from several days after birth until 2 months past term. All infants received a nutrient-enriched (preterm) formula until 2 months past expected term with or without Marine II. Thereafter, all infants received standard term formula.

Table 3. Study II goals

1. To improve AA status in DHA-supplemented infants by feeding a low EPA marine oil and discontinuing DHA supplementation at 2 vs. 9 months after term
2. To improve nutritional status by feeding a nutrient-enriched preterm formula through 2 months after term
3. To monitor ω3 and ω6 status, vitamin A status, visual acuity, mental and psychomotor development and growth

Table 4. Population characteristics

	Study I		Study II	
	control (n = 34)	marine (n = 33)	control (n = 31)	marine (n = 27)
Birth weight, g	1,074	1,147	1,046	1,033
Gestational age, weeks	29.0	29.0	28.0	28.0
Maternal age, years	23	23	24	22
Birth order, no.	2.5	2.3	2.7	2.4
Maternal height, cm	163	165	164	163
Ventilator, h	12.6	6.0	54.6	66.6

Study II Goals

Following completion of study I, we had direct evidence that dietary DHA was conditionally essential for visual development of preterm infants (table 3). However, it was clear that a marine oil source of DHA containing EPA was not the optimal way to improve DHA status. In the course of this study, it became clear that AA status was also poor in formula-fed preterm infants, and poor AA status was associated with poorer growth achievement. Marine oil directly reduced growth achievement and AA status. Moreover, poorer growth achievement was associated with poorer psychomotor development (Bayley Psychomotor Developmental Index (PDI)) at 12 months of age. When study II was designed, one of our goals was to provide DHA while reducing the effects of DHA supplementation on AA status and growth. We achieved this goal by feeding a marine oil source of DHA (Marine II) which contained EPA at a concentration one order of magnitude lower than Marine I and by feeding DHA only through 2 months past term. All infants were provided with a nutrient-enriched preterm formula through 2 months past expected term to ensure that growth and development were not confounded by marginal nutritional status. Repeated assessments of biochemistry, visual acuity, growth and development were made throughout infancy as in study I.

The population characteristics of study I and study II infants are shown in table 4. Compared to study I infants, those in study II were about 1 week younger at birth and required longer periods of mechanical ventilation. The main difference between the two studies was the inclusion of infants with chronic lung disease in study II but not study I. In neither study were infants included who required intestinal surgery or had intraventricular/periventricular hemorrhage > grade 2. Control and experimental infants did not differ in either study.

Preliminary Results of Study II

Although Marine II compared to control infants appear to have better visual acuity as expected at term and 2 months by the Teller Acuity Card procedure, both groups had similar visual acuity from 4 to 12 months. Unlike study I, study II infants supplemented with Marine II did not have poorer AA status, growth achievement or psychomotor development than controls. When Marine II was discontinued at 2 months past expected term, DHA in red blood cell phospholipids began to decline and was not detectably different from controls 4.5 months later. Despite this, infants randomized to Marine II have significantly higher scores on the Bayley Mental Developmental Index (MDI) at 12 months with 90% of the 12-month data completed. We take this as suggestive evidence that very early exposure to DHA is critical for early mental development. The experimental infants in this study were provided DHA throughout the last intrauterine trimester and the first 2 months past expected term.

As shown in table 5, the Bayley MDI did not differ between control and Marine I infants in study I. We can think of two possible explanations why a difference may have been found in study II but not in study I: (1) DHA was provided in a nutrient-enriched formula for approximately 12 weeks longer than in study I. It is possible that marginal nutritional status [19] may have precluded our finding effects of DHA supplementation on early mental development in study I. (2) DHA was fed with very low levels of EPA and stopped before the ω3/ω6 fatty acid balance was disrupted in study II but not in study I. AA and DHA each account for about 25% of the total fatty acids in brain gray matter. Theoretically, this balance could be disrupted by feeding preformed DHA for long periods of time without including AA in the diet, especially since we already know that the AA status of formula-fed preterm infants is marginal compared to term infants [7, 17].

In both study I and study II, we have obtained novelty preference data using the Fagan Infantest procedure. As can be seen in table 6, control and marine oil-supplemented infants appear to have identical means for novelty preference in both studies, suggesting that there is a stable effect of DHA

Table 5. Bayley MDI

Formula at discharge	Treatment	Bayley MDI
Term (study I)	Control	102.1 ± 14.7 (n = 27)
	Marine I	97.4 ± 16.9 (n = 27)
Preterm (study II)	Control	98.0 ± 19.0 (n = 21)
	Marine II	109.7 ± 15.3 (n = 22)[1]

[1] Differs from control discharged on preterm formula ($p < 0.03$).

Table 6. Fagan novelty preference at 12 months

Formula at discharge	Treatment	Novelty preference, %
Term	Control	63.9 ± 7.9 (n = 29)
	Marine I	59.6 ± 6.0 (n = 28)[1]
Preterm	Control	64.9 ± 7.2 (n = 20)
	Marine II	60.9 ± 6.8 (n = 22)[2]

[1] Differs from control discharge on term formula, $p < 0.02$.
[2] Differs from control discharge on preterm formula, $p < 0.075$.

supplementation in both studies. Preliminary analysis of these studies suggests that control and marine oil-supplemented infants have different looking behavior. In study I, Marine I compared to controls had significantly more discrete looks to novel and familiar faces during the novelty portion of the test. In study II, there is a striking positive correlation between the number of discrete looks to novel and familiar faces and the Bayley MDI. The preliminary conclusions from the study are summarized in table 7.

Table 7. Study II conclusions

1. Nutritional status of both dietary groups was improved by discharge on preterm formula
2. Marine II (with 0.03% EPA) improved DHA status as well as Marine I (with 0.3% EPA); plasma and red blood cell DHA declined gradually between 2 and 9 months past term
3. Visual acuity was improved at term and 2 months, but not later in Marine II compared to control infants
4. AA status was better in Marine II compared to Marine I infants
5. Growth and psychomotor development in Marine II infants did not differ from controls
6. Marine II infants have a significantly higher Bayley MDI than controls (109.7 vs. 98.0; $p < 0.03$)
7. Visual attentional style (measured with the Fagan Infantest) in Marine II appears to be similar to Marine I and different from controls in both studies

Table 8. Summary

1. Marine I and Marine II improved early visual acuity
2. Marine II improved scores on a test of early mental development when fed through 2 months past term in conjunction with a nutrient-enriched preterm formula
3. Marine II was not associated with poorer AA status, growth achievement or psychomotor development when discontinued at 2 months past term

Conclusion

Data from two studies provide cause-and-effect evidence that DHA is conditionally essential for early visual acuity development of preterm infants (table 8). These data are in agreement with similar data from Uauy and coworkers [21] who measured visual acuity at 4 months past expected term in control and marine oil-supplemented preterm infants. For the first time, there is also cause-and-effect evidence that early provision of DHA is important for mental development of preterm infants. These latter data come from infants supplemented with a unique, low-EPA marine oil given in conjunction with a nutrient-enriched preterm formula. Finally, it appears that this marine oil can be provided in the first 5 months after birth without the negative effects of AA status, growth achievement or psychomotor development seen in study I.

References

1 O'Brien JS, Fillerup DL, Mean JF: Quantification of fatty acid and fatty aldehyde composition of ethanolamine, choline and serine phosphoglycerides in human gray and white matter. J Lipid Res 1964;5:329–330.

2 Clandinin MT, Chappell JE, Leong S, et al: Intrauterine fatty acid accretion rates in human brain: Implications for fatty acid requirements. Early Hum Dev 1980;4:121–129.

3 Martinez M: Developmental profiles of polyunsaturated fatty acids in the brain of normal infants and patients with peroxisomal diseases: Severe deficiency of docosahexaenoic acid in Zellweger's and pseudo-Zellweger's syndromes; in Simopoulos AP, Kifer RR, Martin RE, et al (eds): Health Effects of Omega-3 Polyunsaturated Fatty Acids in Seafoods. World Rev Nutr Diet. Basel, Karger, 1991, vol 66, pp 87–102.
4 Putnam JC, Carlson SE, DeVoe PW, et al: The effect of variations in dietary fatty acids on the fatty acid composition of erythrocyte phosphatidylcholine and phosphatidylethanolamine in human infants. Am J Clin Nutr 1982;36:106–114.
5 Carlson SE, Rhodes PG, Ferguson MG: Docosahexaenoic acid status of preterm infants at birth and following feeding with human milk or formula. Am J Clin Nutr 1986;44:798–804.
6 Carlson SE, Rhodes PG, Rao V, et al: Effect of fish oil supplementation on the omega-3 fatty acid content of red blood cell membranes in preterm infants. Pediatr Res 1987;21:507–510.
7 Carlson SE, Cooke RJ, Rhodes PG, et al: Long-term feeding of formulas high in linolenic acid and marine oil to very low birth weight infants: Phospholipid fatty acids. Pediatr Res 1991;30:404–412.
8 Farquarson J, Cockburn F, Patrick WA, et al: Infant cerebral cortex phospholipid fatty acid composition and diet. Lancet 1992;340:801–803.
9 Wheeler TG, Benolken RM, Anderson RE: Visual membrane specificity of fatty acid precursors for the electrical response to illumination. Science 1975;188:1312–1314.
10 Neuringer M, Connor WE, Luck SL: Omega-3 fatty acid deficiency in rhesus monkeys: Depletion of retinal docosahexaenoic acid and abnormal electroretinograms. Am J Clin Nutr 1985;43:706.
11 Lamptey MS, Walker BL: A possible essential role of dietary linolenic acid in the development of the young rat. J Nutr 1976;106:86–93.
12 Yamamoto N, Saitoh M, Moriuchi A, et al: Effect of dietary alpha-linolenate/linoleate balance on brain lipid compositions and learning ability in rats. J Lipid Res 1987;28:144–151.
13 Neuringer M, Connor WE, Van Petten, et al: Dietary omega-3 fatty acid deficiency and visual loss in infant rhesus monkeys. J Clin Invest 1984;73:272–276.
14 Carlson S, Cooke R, Werkman S, et al: Docosahexaenoate (DHA) and eicosapentaenoate supplementation of preterm infants: Effects on phospholipid DHA and visual acuity. FASEB J 1989;3:A1056.
15 Carlson SE, Werkman SH, Rhodes PG, et al: Visual acuity development in healthy term infants: Effects of marine oil supplementation. Am J Clin Nutr 1993;58:35–42.
16 Carlson SE, Cooke RJ, Werkman SH, et al: First year growth of preterm infants fed standard compared to marine oil ω3-supplemented formula. Lipids 1992;27:901–907.
17 Carlson SE, Werkman SH, Peeples JM, et al: Arachidonic acid status correlates with first year growth in preterm infants. Proc Natl Acad Sci USA ;1993;90:1073–1077.
18 Carlson SE, Werkman SH, Peeples JM, et al: Growth and development of very low birth weight infants in relation to ω3 and ω6 fatty acid status; in Sinclair A, Gibson R (eds): Essential Fatty Acids and Eicosanoids. Champaign, AOCS Press, 1993, pp 192–196.
19 Peeples JM, Carlson SE, Werkman SH, et al: Vitamin A status of preterm infants during infancy. Am J Clin Nutr 1991;53:1455–1459.
20 Carlson SE, Peeples JM, Werkman SH, et al: Vitamin A status of preterm infants during infancy: Effect of feeding a nutrient-enriched preterm formula after hospitalization. Am J Clin Nutr, submitted.
21 Birch EE, Birch DG, Hoffman DR, et al: Dietary essential fatty acid supply and visual acuity development. Invest Ophthalmol Vis Sci 1992;33:3242–3253.

Susan E. Carlson, PhD, Newborn Center, Room 207, 853 Jefferson Avenue,
Memphis, TN 38163 (USA)

Galli C, Simopoulos AP, Tremoli E (eds): Fatty Acids and Lipids: Biological Aspects.
World Rev Nutr Diet. Basel, Karger, 1994, vol 75, pp 70–78

Polyunsaturated Fatty Acids in the Developing Human Brain, Red Cells and Plasma: Influence of Nutrition and Peroxisomal Disease

Manuela Martinez

Biomedical Research Unit, University Maternity-Children Hospital Valle de Hebron, Barcelona, Spain

In contrast to experimental results in other species [1–3], the fatty acid composition of the developing human brain and retina can be influenced by relatively short periods of nutritional imbalance [4]. It has been previously demonstrated that too high ω6/ω3 ratios may have a strong influence on the polyunsaturated fatty acid (PUFA) composition of tissues when occurring in preterm infants during the vulnerable period of development [5, 6]. The ω3 family is the most affected when an excess of ω6 fatty acids is provided, and docosahexaenoic acid (DHA, 22:6ω3) may be significantly reduced, depending on the timing and duration of the dietary imbalance.

Irrespective of nutrition, children with Zellweger's syndrome (ZS) and related peroxisomal disorders have extremely low levels of DHA in the brain and other tissues [7–9]. In the peroxisomal brain, these low DHA levels are not accompanied by any significant decrease in other PUFAs of the ω6 family. As a consequence, the ω3/ω6 and 22:6ω3/20:4ω6 ratios are strongly reduced, much more so than in normal children fed on extremely high ω6/ω3 intakes. These alterations can be detected in vivo by measuring the PUFA content in red blood cells [10].

A constant finding in peroxisomal disorders is the accumulation of very long chain fatty acids (VLCFA) in tissues and plasma [11, 12], due to defective peroxisomal β-oxidation [13]. According to a recently proposed new metabolic route, DHA is synthesized by β-oxidation of the VLCFA 24:6ω3 [14, 15]. Consequently, it has been suggested that the DHA deficiency in ZS may be directly related to the defective β-oxidation of VLFCAs in this disorder. If so, it should be expected that all patients with peroxisomal β-oxidation defects, including those with late onset peroxisomal disorders, such as X-linked adrenoleukodystrophy (X-ALD) and adrenomyeloneuropathy (AMN), have low levels of DHA.

This paper summarizes the most interesting PUFA changes in the normal human developing brain and in patients with generalized peroxisomal (GP) disorders (due to defective peroxisome biogenesis). As an approximate in vivo indicator of brain PUFA status, the main red cell PUFA data in 51 patients with different peroxisomal disorders are presented. The results are compared with those in 33 normal controls fed on balanced Western diets and with 4 strictly vegetarian normal children. Some conclusions are drawn as for the differences between X-ALD/AMN and GP patients, and the influence of nutrition on both groups of patients, as well as on the normal developing human being.

Materials and Methods

The brain study included 16 patients with GP disorders, who had been diagnosed of either ZS or neonatal adrenoleukodystrophy (NALD), plus 1 patient with isolated bifunctional enzyme deficiency (BED). The age of these patients ranged between 3 months and 3 years. Blood samples from 24 other patients with GP disorders were collected (ages comprised between 2 months and 19 years). For comparison, blood samples were also drawn from 27 patients with X-ALD or AMN, aged between 5 and 44 years, from 33 omnivorous, healthy subjects (ages 3–50 years) and from 4 Indian, strictly vegetarian, normal children (ages 4–12 years).

Total fatty acid methyl esters (FAMEs) were obtained by direct transesterification [16] of brain tissue, red cells and plasma with HCl-methanol, as specified elsewhere [8]. Separation was effected on a 30-m long, 0.25 mm ID SP-2330 column, installed in a Hewlett-Packard 5890 gas chromatograph, and operated on programmed temperature (140–200 °C at 3 °C/min). Peaks were identified by comparison of the retention times with those of pure standards and by mass spectrometry of methyl esters and/or picolinyl esters [17], with a Hewlett-Packard 5970B mass selective detector.

Results and Discussion

DHA and AA in the Brain of Normal and Peroxisomal Children

As figure 1 shows, DHA concentration increases steadily throughout development until about 2 years of age. This increase is most rapid during the last trimester of intrauterine life [4]. When there is nutritional imbalance, occurring during this rapid developmental period, due to diets with very high ω6/ω3 ratios, a DHA reduction occurs both in the brain and retina [4–6]. Apart

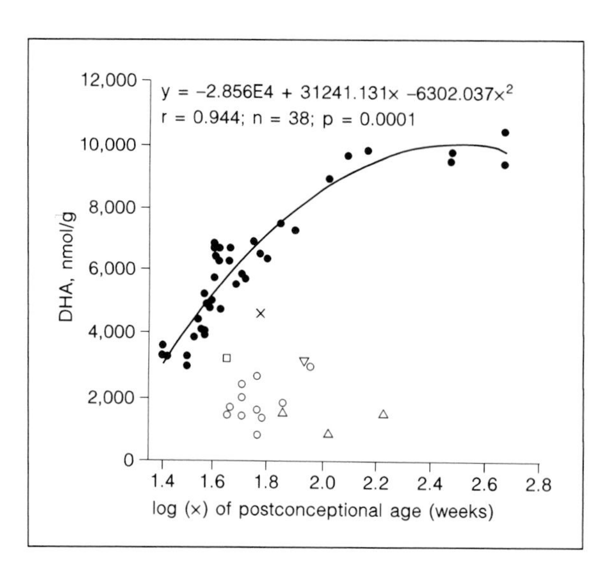

Fig. 1. When plotted against the logarithm of postconceptional age (in weeks), the concentration of DHA in the normal developing brain (●) increases in a fast, parabolic manner until about 2 years of age (log, 2.16). □ = Malnourished child on very high ω6/ω3 diets; ○ = ZS; △ = NALD; ▽ = pseudo-ZS; × = BED. Notice that all children with GP disorders have very low brain DHA values. Both the patient with BED and the malnourished child have more discretely reduced brain DHA concentrations.

from dietary factors, children with GP disorders have very low tissue concentrations of DHA [7–9]. Figure 1 summarizes these data in the brain and shows that ZS, as well as NALD and psuedo-ZS, all result in very low brain DHA values. A case with isolated deficiency of bifunctional enzyme had a brain DHA concentration much less reduced than the GP patients.

The concentration of arachidonic acid (AA, 20:4ω6) also increases very fast in the developing brain up until about 2 years of age (fig. 2). Since most milk formulas are mainly enriched in linoleic acid (LA, 18:2ω6), the ω6/ω3 ratio is usually high, and the AA concentration is expected to be maintained in the brain, as happened with the malnourished child in figure 2. In contrast to DHA, figure 2 shows that the concentration of AA was high or normal in the brain of children with ZS. Only 1 patient with NALD and another with pseudo-ZS had somewhat low brain AA levels.

PUFAs in Red Cells and Plasma in Patients with Different Peroxisomal Disorders

The deficiency of DHA in peroxisomal tissues is not accompanied by any significant decrease in other ω3 fatty acids, and the precursor, 22:5ω3, is either

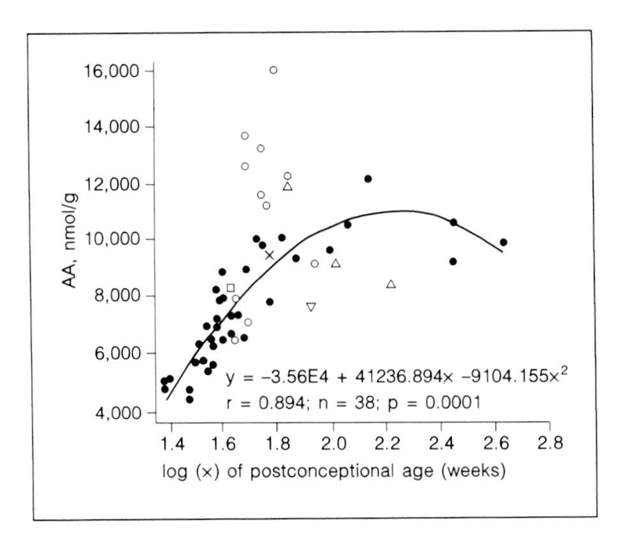

Fig. 2. Like DHA, AA increases in the normal human brain until about 2 years of age. Symbols as in figure 1. ZS patients have high to normal brain AA values.

within normal limits or even somewhat increased. As a consequence, a constant finding in GP patients is an important decrease of the ratio 22:6ω3/22:5ω3. In red cells, this ratio very often becomes lower than 1, which is an inversion of the normal situation, where the 22:6ω3/22:5ω3 ratio is higher than 2. In order to separate the effects of nutrition from those of peroxisomal disease, figure 3 shows both PUFAs, 22:5ω3 and 22:6ω3, in red cells from differently nourished controls and peroxisomal patients. It can be seen that vegetarian controls had DHA levels within the same low range as GP patients, and 22:6ω3/22:5ω3 ratios almost as low. In contrast, when X-ALD/AMN patients receiving a complete diet (with the exclusion of saturated fat) are compared with omnivorous controls, figure 3 shows that both 22:6ω3 and 22:5ω3 were normal in red cells. However, 2 X-ALD patients fed on elemental diets by intubation had red cell DHA and 22:6ω3/22:5ω3 ratios as low as GP patients. Similar findings were dectected in plasma (fig. 4).

Irrespective of the diet, AA was within normal limits in red cells from generalized peroxisomal, as well as from X-ALD/AMN patients and strictly vegetarian children (fig. 5). LA was low in the 2 X-ALD patients on elemental diets and high in GP patients and vegetarian controls. An important difference between the latter two groups of children, however, was that the 20:4ω6/18:2ω6 ratio was low in GP patients only. In plasma (fig. 6), LA was very low in 1 of the X-ALD patients nourished on elemental diets. This child was a strict vegetarian and was nourished on very poor PUFA diets. However, his AA levels were good.

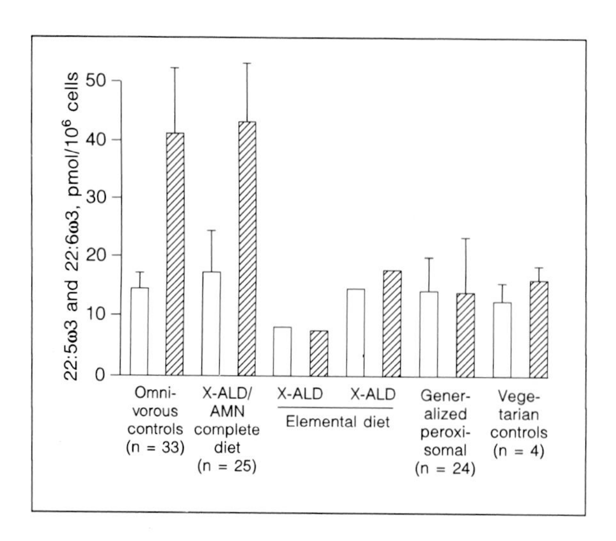

Fig. 3. Grouped data are means ± 1 SD. ▨ = 22:6ω3 (DHA); □ = 22:5ω3. Children with GP disorders, vegetarians and patients with poor PUFA intake all have low levels of DHA in red cells.

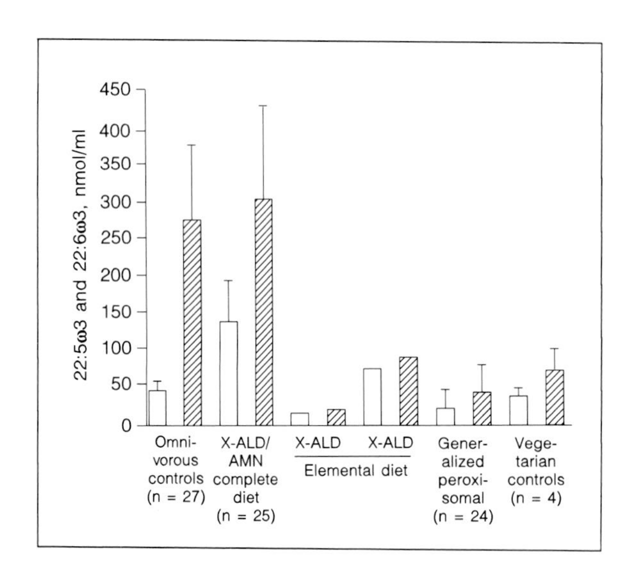

Fig. 4. 22:5ω3 and 22:6ω3 in plasma from peroxisomal patients and controls. Symbols as in figure 3.

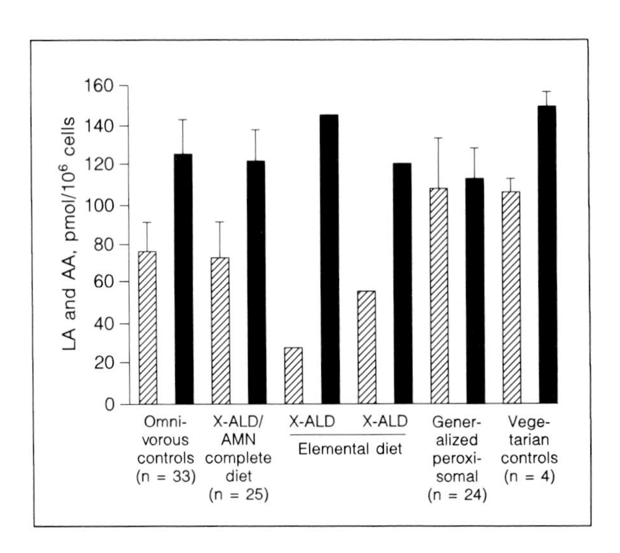

Fig. 5. LA (▨) and AA (■) in red cells from peroxisomal patients and controls.

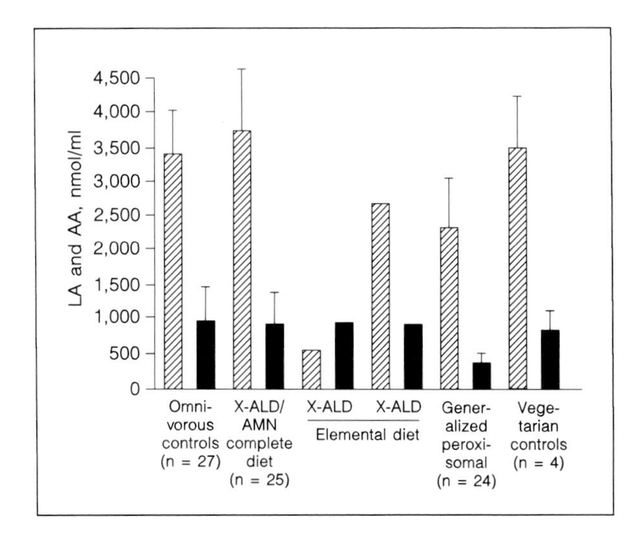

Fig. 6. Plasma AA is normal in all cases except in patients with generalized disorders. Symbols as in figure 5.

An unexpected finding in plasma from GP patients was the significantly low levels of AA, not matched by any of the other peroxisomal or control groups. The low AA values in plasma of GP patients, together with the high tissue levels of LA and the low red cell 20:4ω6/18:2ω6 ratio may suggest a deficient conversion of linoleate into arachidonate in children with generalized peroxisomal disorders, possibly due to a deficient Δ6 and/or Δ5 desaturase. Simple nutritional deficiency in these children, usually maintained on low fat diets, could be an alternative explanation, consistent with the low plasma levels of LA found in GP patients.

Conclusions

It can be concluded that the DHA deficiency is a constant finding only in generalized peroxisomal disorders. On the other hand, under good nutritional conditions, X-ALD/AMN patients have normal levels of DHA, as well as of other PUFAs, at least in red cells. If DHA is produced by β-oxidation of 24:6ω3, it is hard to explain why some of the patients with impaired β-oxidation of VLCFAs, as X-ALD/AMN patients, should have normal DHA synthesis. Whatever the metabolic route for the synthesis of long chain PUFAs, the present results suggest that the pathogenetic mechanism for early and late onset peroxisomal disorders must be quite different.

These data show that nutrition plays an important role in altering blood DHA levels and the 22:6ω3/22:5ω3 ratio. This happens in normal individuals, as well as in patients with peroxisomal disease, and emphasizes the importance of comparing the data obtained in patients with those in controls on similar diets. So it seems that both nutrition and peroxisomal disease can reduce the DHA levels significantly. The difference is that the DHA deficiency in peroxisomal patients is produced even with normal ω3 intakes, and affects brain composition and function.

The extreme DHA deficiency in the peroxisomal brain is, for the moment, difficult to explain. DHA is widely distributed in food and very little seems to be enough for normal brain development [18]. It is, therefore, not at all obvious why peroxisomal patients should have a DHA deficiency of such a considerable magnitude. If local synthesis in the brain is the main problem, one does not understand why the DHA deficiency should be also reflected in red cells and, if so, why these cells incorporate so very little of the DHA from the food. On the other hand, it has been demonstrated that DHA administered orally as ethyl ester is readily absorbed and incorporated into erythrocyte membranes until normalization of their DHA content [19]. This has produced some clear signs of neurological improvement that suggest incorporation of DHA into the brain also. So, in a way, the peroxisomal DHA deficiency behaves like a nutritional deficiency. In this context, it is interesting to point out that some generalized peroxisomal patients have very high brain levels of 22:5ω6 [9], as happens in nutritional ω3 deprivation [20, 21]. The obvious conclusion is that DHA should be provided to all peroxisomal patients with a clear DHA deficiency.

The low values of plasma arachidonate in patients with generalized peroxisomal disorders were surprising considering the high levels of this PUFA in the brain of ZS patients and the normal levels in red blood cells. The low concentration of AA in plasma could indeed be of nutritional origin, since these patients are usually subjected to strict diets, very low in fat and virtually devoided of PUFAs. However, considering that an increase in the parent fatty acid, linoleate, was a quite constant finding in peroxisomal tissues and that the ratio 20:4ω6/18:2ω6 was decreased, even in the peroxisomal brain, the existence of a deficiency of Δ5 and/or Δ6 desaturation cannot be ruled out. In any case, it is tempting to speculate that such low plasma AA levels could be related to the poor growth in these patients, as has been shown to happen in the normal preterm infant during early development [22]. Normalization of plasma arachidonate is recommended, together with the DHA therapy, especially since an increase in ω3 fatty acids could secondarily cause a further reduction in ω6 PUFAs.

Acknowledgements

I thank the following investigators for kindly providing the peroxisomal samples reported in this study: Dr. I. Lorente (Hospital de Sabadell, Barcelona), Dr. M. Pineda (Hospital de San Juan de Dios, Barcelona), Dr. R. Vidal (Hospítal de Tarrasa, Barcelona), Dr. M. Giros, and Dr. T. Pampols (Instituto de Bioquímica Clínica, Barcelona), Dr. G. Lorenzo (Hospital Ramón y Cajal, Madrid), Dr. J.M. Prats (Hospital de Cruces, Bilbao), Dr. H.W. Moser and Ann Moser (The Kennedy Institute, Baltimore), Dr. Frank Roels (Universiteit Gent), Dr. P.G. Barth, Dr. R.J.A. Wanders and Dr. R.B.H. Schutgens (University Hospital, Amsterdam), Dr. J.M. Saudubray (Hôpital Necker des Enfants Malades, Paris), Dr. H. Mandel (Rambam Medical Center, Haifa), Dr. H. Souayah and Dr. C.A. Peltier (Hôpital Universitaire Saint-Perre, Brussels), Dr. P.T. Clayton, (Institute of Child Health, London), Dr. L. Gopolakrishnan (Madras), Dr. L. van Maldergem (Institut de Morphologie Pathologique, Loverval), Dr. J. van Moorsel (Free University Hospital, Amsterdam), Dr. Ph. Evrard (Cliniques Universitaires Saint-Luc, Brussels), Dr. E. Vamos (Hôpital Universitaire Brugmann, Brussels), Dr. G. Ferriere, Dr. Ch. Bonnier, and Dr. A. Witgens (Centre Neurologique William Lennox, Louvain-la-Neuve), Dr. F. Hanefeld, and Dr. C. Korenke (Georg-August-Universität, Göttingen), Dr. A. Nogueira (Instituto de Genética Médica Jacinto de Magalhães, Porto), Dr. W. Kohler (Krankenhaus Moabit, Berlin) and Mr. Axel Wrede (Informationsnetzwerk Leukodystrophie, Bonn).

References

1 Mohrhauer H, Holman RT: Alteration of the fatty acid composition of brain lipids by varying levels of dietary essential fatty acids. J Neurochem 1963;10:523–530.
2 Galli C, Trzeciak HI, Paoletti R: Effects of dietary fatty acids on the fatty acid composition of brain ethanolamine phosphoglyceride: Reciprocal replacement of ω6 and ω3 polyunsaturated fatty acids. Biochim Biophys Acta 1971;248:449–454.

3 Anderson RE, Maude MB: Lipids of ocular tissues. VIII. The effects of essential fatty acid deficiency on the phospholipids of the photoreceptor membranes of rat retina. Arch Biochem Biophys 1971;151:270–276.
4 Martinez M: Tissue levels of polyunsaturated fatty acids during early human development. J Pediatr 1992;120:S129–S138.
5 Martinez M, Ballabriga A: Effects of parenteral nutrition with high doses of linoleate on the developing human liver and brain. Lipids 1987;22:133–138.
6 Martinez M, Ballabriga A, Gil-Gibernau JJ: Lipids of the developing human retina. I. Total fatty acids, plasmalogens, and fatty acid composition of ethanolamine and choline phosphoglycerides. J Neurosci Res 1988:20:484–490.
7 Martinez M: Polyunsaturated fatty acid changes suggesting a new enzymatic defect in Zellweger syndrome. Lipids 1989;24:261–265.
8 Martinez M: Severe deficiency of docosahexaenoic acid in peroxisomal disorders: A defect of Δ4-desaturation? Neurology 1990;40:1292–1298.
9 Martinez M: Abnormal profiles of polyunsaturated fatty acids in the brain, liver, kidney and retina of patients with peroxisomal disorders. Brain Res 1992;583:171–182.
10 Martinez M: Treatment with docosahexaenoic acid favorably modifies the fatty acid composition of erythrocytes in peroxisomal patients; in Coates PM, Tanaka K (ed): New Developments in Fatty Acid Oxidation. New York, Wiley-Liss, 1992, pp 389–397.
11 Igarashi M, Schaumburg HH, Powers J, Kishimoto Y, Kolodny E, Suzuki K: Fatty acid abnormality in adrenoleukodystrophy. J Neurochem 1976;25:851–860.
12 Brown FR, McAdams AJ, Cumins JW, Konkol R, Singh I, Moser AB, Moser HW: Cerebro-hepato-renal (Zellweger) syndrome and neonatal adrenoleukodystrophy: Similarities in phenotype and accumulation of very long chain fatty acids. Johns Hopkins Med J 1982;151:344–361.
13 Singh I, Moser AB, Goldfischer S, Moser HW: Lignoceric acid is oxidized in the peroxisome: Implications for the Zellweger cerebro-hepato-renal syndrome and adrenoleukodystrophy. Proc Natl Acad Sci USA 1984;81:4203–4207.
14 Rosenthal MD, Garcia MC, Jones MR, Sprecher H: Retroconversion and Δ^4-desaturation of docosatetraenoate (22:4(ω6)) and docosapentaenoate (22:5(ω3)) by human cells in culture. Biochim Biophys Acta 1991;1083:29–36.
15 Voss A, Reinhart M, Sankarappa S, Sprecher H: The metabolism of 7,10,13,16,19-docosapentaenoic acid to 4,7,10,13,16,19-docosahexaenoic acid in rat liver is independent of a 4-desaturase. J Biol Chem 1991;266:19995–20000.
16 Lepage G, Roy CC: Direct transesterification of all classes of lipids in a one-step reaction. J Lipid Res 1986;27:114–120.
17 Christie WW, Brechany EY, Johnson SB, Holman RT: A comparison of pyrrolidide and picolinyl ester derivatives for the identification of fatty acids in natural samples by gas chromatography-mass spectrometry. Lipids 1986;21:657–661.
18 Crawford MA, Sinclair AJ, Msuya PM, Murhambo A: Structural lipids and their polyenoic constituents in human milk; in Galli C, Jacini G, Pecile A (eds): Dietary Lipids and Postnatal Development. New York, Raven Press, 1973, pp 41–56.
19 Martínez M, Pineda M, Vidal R, Conill J, Martin B: Docosahexaenoic acid: A new therapeutic approach to peroxisomal-disorder patients. Experience with two cases. Neurology 1993;43:1389–1397.
20 Neuringer M, Connor WE, Lin DS, Barstad L, Luck S: Biochemical and functional effects of prenatal and postnatal ω3 fatty acid deficiency on retina and brain in rhesus monkeys. Proc Natl Acad Sci USA 1986;83:4021–4025.
21 Bourre JM, Francois M, Youyou A, Dumont O, Piciotti M, Pascal G, Durand G: The effects of dietary α-linolenic acid on the composition of nerve membranes, enzymatic activity, amplitude of electrophysiological parameters, resistance to poisons and performance of learning tasks in rats. J Nutr 1989;119:1880–1892.
22 Carlson SE, Werkman SH, Peeples JM, Cooke RJ, Tolley EA: Arachidonic acid status correlates with first year growth in preterm infants. Proc Natl Acad Sci USA 1993;90:1073–1077.

Manuela Martinez, MD, Biomedical Research Unit, University Maternity-Children Hospital Valle de Hebron, E-08035 Barcelona (Spain)

Galli C, Simopoulos AP, Tremoli E (eds): Fatty Acids and Lipids: Biological Aspects.
World Rev Nutr Diet. Basel, Karger, 1994, vol 75, pp 79–81

Summary Statement: Maternal and Infant Nutrition

Berthold Koletzko

The session was co-chaired by *B. Koletzko* and *R. Uauy*, and presentations were made by Drs. *Koletzko, M. Hamosh, E. Lien, G. Crozier, T.A.B. Sanders, P. Budowski* and *A.D. Postle.*

Rather than repeating the contents of the abstracts of papers presented at the session on maternal and infant nutrition, it is tried here to present some general conclusions that can be drawn at this time with a reasonable degree of confidence.

With respect to nutrition of pregnant women, there is a large body of evidence indicating an association of poor socioeconomic living conditions, poor quality of maternal diet, and poor outcome of pregnancy with low average birthweights and high perinatal and infant morbidity. Normal intrauterine growth and development depend on adequate availability and metabolism of nutrients. An impressive example is the documented deleterious effect of poor maternal folic acid supply that results in a high incidence of infantile neural tube defects. Among other substrates, the fetus requires for deposition in its rapidly growing tissues appreciable amounts of polyunsaturated fatty acids (PUFA), and their availability is related to maternal diet and metabolism. Several studies showed marked effects of maternal diet and disease on fatty acid metabolism in the fetus, and there is certainly a need for further research in this important area.

Several pieces of evidence discussed at this conference indicate an essential role of docosahexaenoic acid (DHA) for early human development. In utero, the placenta appears to supply DHA and other long chain polyunsaturated fatty acids (LCPUFA) to the fetus. After birth, breast-feeding is considered the best choice of feeding for a healthy infant to provide an adequate nutrient supply. Human milk supplies significant amounts of ω3 and ω6 LCPUFA, including

DHA and arachidonic acid (AA). In contrast, most of today's artificial feeding regimens for young infants are devoid of appreciable amounts of LCPUFA. In premature infants raised on formula providing linoleic (LA) and α-linolenic acids (LNA) but none of their products, signs of functional disadvantage have been found which could be corrected by a dietary supply of preformed DHA. Also, dietary regimens which lead to a depletion of AA were associated with functional disadvantages. Obviously, there is an urgent need to improve the PUFA supply for infants born prematurely in order to prevent a state of deficiency.

Intensive research is presently being conducted to evaluate the proposed different practical approaches to improve infants' PUFA status and to define the optimal lipid composition of the diet, their short- and long-term effects, and their safety, but this should not prevent us from aiming at the provision of an improved nutrient supply to this high-risk group as soon as possible. In Europe, the Committee on Nutrition of the European Society for Paediatric Gastroenterology and Nutrition (ESPGAN) in 1991 recommended to enrich formula for low birthweight infants with metabolites of both LA and LNA. Since the exact requirements of premature infants are not yet known, it was recommended that the average composition of human milk should be used as a model for the composition of low birthweight infant formula until better data on optimal intakes become available.

Furthermore, it was discussed at this meeting that feeding vegetable oil-based formula to healthy infants born at full term also leads to a depletion of DHA and AA. There are indications that this depletion may be associated with functional disadvantages also in full-term infants, but these data await further confirmation. However, there is general agreement that all infant diets should contain at least LNA as a source of ω3 fatty acids, and it is considered unacceptable that there are some infant foods still in use today in some parts of the world that provide practically no ω3 fatty acids. It was also agreed that one should avoid extreme ratios of LA and LNA in infant formula. The range of 5–15/1 for the LA/LNA ratio in infant formula recommended by the ESPGAN Committee on Nutrition was considered reasonable.

In the past, the present level of dietary intake of *trans* isomeric fatty acids during pregnancy, lactation and infancy was regarded safe. However, data presented at this conference demonstrate that high prenatal exposure to *trans* isomeric fatty acids is linked to poor availability of LCPUFA and lower birthweight in human infants, which is in agreement with previous observations in animal studies. These results question the safety of a high dietary supply of *trans* fatty acids in the perinatal period and add further weight to the demand for a reduction of the high trans fatty acid contents in the current diets in many countries. As a first step in this direction, a regulation has been proposed for the European Community to restrict the maximum content of *trans* fatty acids in infant formulas to no more than 4% of total fat.

Finally, it is proposed that we try harder to use one language. With the increased relevance of lipid biochemistry for the practice of nutrition and clinical medicine, a common and generally accepted terminology becomes much more important. It does not matter much for specialists in the field if several terms are used simultaneously for one and the same fatty acid, but this is very confusing for those not familiar with the area, such as most clinicians. We need to get relevant messages across to obstetricians, pediatricians, nurses and many others. For this purpose it is unfortunate that numerous terms are used for the same group of fatty acids, including essential fatty acid metabolites, product essential fatty acids, desaturation/elongation products, HUFA (highly unsaturated fatty acids), LCP, LC-PUFA and LCPUFA (long chain polyunsaturated fatty acids), VLCP, VLC-PUFA and VLCPUFA (very long chain polyunsaturated fatty acids). I sincerely hope that the International Society for the Study of Fatty Acids and Lipids (ISSFAL) can contribute to an agreement on a common terminology. It does not really matter which terms are used, as long as most of us agree on one. However, I must admit that I favor the term 'LCP' for the above-mentioned group of fatty acids, because this term was coined already some two decades ago by Drs. Andy Sinclair and Michael Crawford, two highly respected pioneers in the field, and it has also been adopted in documents of official bodies such as the European Community.

Galli C, Simopoulos AP, Tremoli E (eds): Fatty Acids and Lipids: Biological Aspects.
World Rev Nutr Diet. Basel, Karger, 1994, vol 75, pp 82–85

Trans Fatty Acids and the Human Infant[1]

Berthold Koletzko

Kinderpoliklinik der Ludwig-Maximilians-Universität München, Germany

The human diet contains large amounts of *trans* fatty acids that originate primarily from partially hydrogenated fats. The consumption of *trans* fatty acids has been considered safe for humans by expert committees reporting to the US Food and Drug Administration and to the British Nutrition Foundation [1, 2], but this conclusion appears questionable in the light of recent studies indicating possible untoward effects on lipoprotein metabolism and risk of coronary heart disease in human adults [3, 4]. With respect to the perinatal period, the American consensus report also concluded that the human placenta acts as a barrier for *trans* fatty acids and hence excluded a significant exposure and untoward effects for the fetus [1]. We rejected this conclusion when we detected *trans* fatty acids at similar percentage levels in cord and in maternal plasma lipids of 30 pairs of mothers and their term infants at the time of birth [5]. These data provide evidence for a placental transfer of *trans* fatty acids that are not synthesized by the human fetus. This finding prompted us to study indicators of potential adverse effects of perinatal *trans* isomer exposure.

Previous studies in rats in vivo and in rodent tissues and human fibroblasts in vitro demonstrated that *trans* fatty acids impair microsomal desaturation and chain elongation of linoleic (18:2ω6 LA), and α-linolenic acids (18:3ω3 LNA) to their long-chain polyunsaturated fatty acid (LCPUFA) metabolites with 20 and 22 carbon atoms [reviewed in 6]. LCPUFA such as arachidonic (20:4ω6 AA) and docosahexaenoic acids (22:6ω3 DHA) are required in appreciable amounts during perinatal development as essential membrane components and as eicosanoid precursors [7]. Moreover, newborn rats and

[1] The work of the author is financially supported by Deutsche Forschungsgemeinschaft, Bonn, Germany (Ko 912/4-2).

Table 1. Significant linear correlation coefficients (r) between weight percentage values of *trans* fatty acids (elaidic acid, 18:1t and total *trans* fatty acids) and LCPUFA or the product/substrate ratios for LCPUFA biosynthesis in plasma lipid classes of 29 premature infants on day 4 of life (* $p<0.05$ and ** $p<0.01$)

	Triglycerides		Sterolesters		Phospholipids	
	18:1t/total	*trans*	18:1t/total	*trans*	18:1t/total	*trans*
18:2ω6 (linoleic)	n.s.	n.s.	n.s.	n.s.	n.s.	n.s.
20:4ω6 (arachidonic)	n.s.	n.s.	−0.45*	−0.38*	n.s.	n.s.
Total ω6 LCPUFA	−0.41*	−0.47*	−0.43*	−0.41*	n.s.	−0.40*
18:3ω3 (α-linolenic)	n.s.	n.s.	n.s.	n.s.	n.s.	n.s.
22:6ω3 (docosahexaenoic)	−0.44*	−0.41*	−0.41*	−0.53**	n.s.	n.s.
Total ω3 LCPUFA	−0.50*	−0.38	n.s.	n.s.	n.s.	n.s.
Total ω6 + ω3 LCPUFA	−0.55**	−0.51**	n.s.	−0.40*	n.s.	n.s.
Ratio 20:4ω6/18:2ω6	−0.47**	−0.41*	−0.46*	n.s.	n.s.	n.s.
Ratio 22:6ω3/18:3ω3	−0.50**	−0.48**	−0.42*	−0.46*	n.s.	n.s.

mice fed *trans* fatty acids showed impaired postnatal weight gain, and a possible reduction of intrauterine growth was also suggested by a trend to lower birthweights in the offspring of sows fed partially hydrogenated oils [reviewed in 6].

In view of these documented risks of perinatal *trans* fatty acid exposure in animals, we investigated the potential of similar side effects on LCPUFA status and body weight in a group of 29 premature infants with a birthweight of 1,700 ± 127 g (mean ± SD) and gestational age of 33.6 ± 1.4 weeks [6]. The protocol of the study was reviewed by the local ethics committee, and informed parental consent was obtained. Fatty acid composition of plasma lipids was determined within 96 h after birth as previously described, with a precision for determination of elaidic acid (18:1t) and total *trans* fatty acids of 5.0 and 3.8%, respectively [8–10].

We found *trans* fatty acids in plasma triglycerides, sterolesters and phospholipids of every infant. At the time when these infants were studied, they had only just started enteral feeding and a catabolic state prevailed, in which fatty acid composition of their plasma lipids reflects primarily intrauterine nutrient supply and metabolism. In agreement with our previous observations in premature babies, healthy infants and young children [11], *trans* fatty acids were not randomly distributed within the three main plasma lipid classes. Rather, we found a relatively low contribution of elaidic acid (18:1t) to total *trans* isomers in sterolesters suggestive of a low 18:1t esterification with cholesterol [6].

Table 2. Significant linear correlation coefficients (r) for birthweight (g) and % wt/wt values of *trans* fatty acids (elaidic acid [18:1t] and total *trans* fatty acids) in plasma sterolesters and phospholipids of 29 premature infants on day 4 of life (*$p<0.05$ and **$p<0.01$)

	Sterolesters	Phospholipids
18:1t (elaidic)	-0.50**	n.s.
Total *trans*	-0.40*	-0.42**

Linear regression analysis demonstrated no correlation between *trans* fatty acids and the substrates of LCPUFA biosynthesis, LA and LNA (table 1). In contrast, there was an inverse correlation of percentage values of elaidic acid (18:1t) and total *trans* isomers with LCPUFA content, particularly in triglycerides and sterolesters (table 1). In both lipid classes, elaidic and total *trans* fatty acids were also inversely correlated to the AA/LA and DHA/LNA ratios, i.e. the product/substrate ratios for ω6 and ω3 LCPUFA biosynthesis. Total *trans* isomers in sterolesters and phospholipids and elaidic acid in sterolesters were also negatively correlated to infantile birthweight (table 2), but not to gestational age.

Our observations are compatible with a *trans* fatty acid – induced impairment of desaturation and/or chain elongation of essential fatty acids in humans during the perinatal period. In animal studies and in vitro experiments, both Δ-6- and Δ-5-desaturases are suppressed by *trans* isomers of LA, but also by elaidic acid (18:1t) [reviewed in 6]. During fetal and early postnatal development, ω6 and ω3 LCPUFA are required in large amounts for the deposition in membrane lipids of the fast growing brain, retina and other organs [7, 12], but endogenous LCPUFA synthesis is limited [7, 10]. Since the quality of early LCPUFA accretion in brain and retina is related to the development of neural and visual functions in newborn animals and in premature infants [7, 13, 14], a further inhibition of the low capacity of LCPUFA synthesis in the perinatal period by *trans* fatty acids could have serious effects on functional development.

Our findings indicate growth impairment as another potential risk of a high perinatal *trans* fatty acid exposure. In newborn rats and mice, postnatal growth was impaired by *trans* isomers of LA and by partially hydrogenated soybean oil. Growth failure did not occur when very high amounts of essential fatty acids (EFA) were supplied along with the *trans* isomers [15]. However, newborn and especially premature infants have extremely limited reserves of EFA and, therefore, may not be able to compensate for distortions of EFA metabolism induced by *trans* isomers.

Intrauterine *trans* fatty acid exposure is expected to depend primarily on maternal consumption of hydrogenated fats, just as *trans* fatty acid concentration in human milk depends on maternal diet [9, 11]. *Trans* fatty acids are also found in some infant formulas, presumably related to the use of butterfat and other animal fat for formula production [16]. The data presented here raise serious concerns as to the safety of dietary *trans* fatty acids in pregnant and lactating women and in newborn infants, and they should prompt further detailed investigation of potential adverse effects.

References

1 Senti FR (ed): Health aspects of dietary *trans* fatty acids. Report prepared for Center for Food Safety and Applied Nutrition, Food and Drug Administration, Department of Health and Human Services, Washington, DC 20204. Bethesda, Life Sciences Research Office, Federation of the American Societies for Experimental Biology, 1985.
2 British Nutrition Foundation Task Force: Report on *trans* fatty acids. London, The British Nutrition Foundation, 1987.
3 Zock PL, Katan MB: Hydrogenation alternatives: Effects of trans fatty acids and stearic acid versus linoleic acid on serum lipids and lipoproteins in humans. J Lipid Res 1992;33:399–410.
4 Willett WC, Stampfer MJ, Manson JE, Colditz GA, Speizer FE, Rosner BA, Sampson LA, Hennekens CH: Intake of trans fatty acids and risk of coronary heart disease among women. Lancet 1993;341:581–585.
5 Koletzko B, Müller J: *Cis*- and *trans*-isomeric fatty acids in plasma lipids of newborn infants and their mothers. Biol Neonate 1990;57:172–178.
6 Koletzko B: Trans fatty acids may impair biosynthesis of long-chain polyunsaturates and growth in man. Acta Paediatr 1992;81:302–306.
7 Koletzko B: Fats for brains. Eur J Clin Nutr 1992;46:S51–62.
8 Koletzko B, Mrotzek M, Bremer HJ: Fatty acid composition of mature human milk in Germany. Am J Clin Nutr 1988;47:954–959.
9 Koletzko B: Zufuhr, Stoffwechsel und biologische Wirkungen *trans*-isomerer Fettsäuren bei Säuglingen. Nahrung 1991;35:229.
10 Koletzko B, Schmidt E, Bremer HJ, Haug M, Harzer G: Effects of dietary long-chain polyunsaturated fatty acids on the essential fatty acid status of premature infants. Eur J Pediatr 1989;148:669–675.
11 Koletzko B, Mrotzek M, Bremer HJ: *Trans* fatty acids in human milk and infant plasma and tissue; in Goldman AS, Atkinson S, Hanson LA (eds): Human Lactation, vol 3: Effect of Human Milk on the Recipient Infant. New York, Plenum Publishing, 1987, pp 323–333.
12 Koletzko B: Essentielle Fettsäuren: Bedeutung für Medizin und Ernährung. Aktuel Endokrinol Stoffwechsel 1986;7:18–27.
13 Carlson SE, Cooke RJ, Peeples JM, Werkman SH, Tolley EA: Docosahexaenoate and eicosapentanoate status of preterm infants: Relationship to visual acuity in n-3 supplemented and unsupplemented infants. Pediatr Res 1989;25:285A.
14 Uauy R: Are ω-3 fatty acids required for normal eye and brain development in the human? J Pediatr Gastroenterol Nutr 1990;11:296–302.
15 Alfin-Slater RB, Wells P, Aftergood L: Dietary fat composition and tocopherol requirement. IV. Safety of polyunsaturated fats. J Am Oil Chem Soc 1973;50:479–484.
16 Koletzko B, Bremer HJ: Fat content and fatty acid composition of infant formulae. Acta Paediatr Scand 1989;78:513–521.

Prof. Dr. med. Berthold Koletzko, Kinderpoliklinik, Klinikum Innerstadt der Ludwig-Maximilians-Universität, Pettenkoferstrasse 8a, D-80336 München (FRG)

Galli C, Simopoulos AP, Tremoli E (eds): Fatty Acids and Lipids: Biological Aspects.
World Rev Nutr Diet. Basel, Karger, 1994, vol 75, pp 86–91

Milk Lipids and Neonatal Fat Digestion: Relationship between Fatty Acid Composition, Endogenous and Exogenous Digestive Enzymes and Digestion of Milk Fat

Margit Hamosh, Sara J. Iverson, Charlotte L. Kirk, Paul Hamosh

Departments of Pediatrics and Physiology and Biophysics, Georgetown University Medical Center, Washington, D.C., USA

A major shift occurs at birth from the predominantly carbohydrate-rich nutrition of the fetus to the high fat diet provided by mother's milk or formula. Fat is, however, not only the major energy source of the infant, it also provides specific components with well-defined functions. Among these are medium chain fatty acids (MCFA) which can be absorbed directly from the stomach [1] and therefore provide a more readily accessible energy source than carbohydrate, and long chain polyunsaturated fatty acids (LCPUFA) which are essential for brain development and retinal function, as well as being precursors for prostaglandins and cytokines. The accessibility of these fatty acids to the infant depends upon their presence in the diet as well as the efficiency of fat digestion and absorption. In contrast to human milk, infant formula lacks LCPUFA. The premature infant is born with very low LCPUFA reserves (both ω3 and ω6) because fetal fat deposition and storage occur only during the last trimester of gestation. The need for these fatty acids is met by an optimal balance between ω3 and ω6 LCPUFA in milk [2, 3]; however, attempts to achieve such a balance in formulas are still in the experimental stage.

While fat absorption seems to be adequate in the newborn [4], the intestinal phase of fat digestion is deficient because of low pancreatic lipase and bile salt levels. Adequate assimilation of lipid depends therefore upon alternate digestive mechanisms, such as lipolysis in the stomach by gastric lipase followed by intestinal hydrolysis by a specific milk digestive lipase.

Table 1. Fat content and fatty acid composition of human, dog and ferret milk

	Human [1]	Dog [2]	Ferret
Total Fat, g/dl	3.60	7.90	6.70
Fatty acid, wt%			
8:0–12:0	5.43	0.17	0.36
14:0	5.68	3.25	3.80
16:0–18:0	29.90	32.20	31.80
18:1	35.51	40.72	29.45
18:2ω6	15.58	11.92	12.76
18:3ω3	1.03	0.49	0.06
20:4ω6	0.60	1.15	0.52
22:5ω3 + 22:6ω3	0.34	0	0.63

[1] Data from Bitman et al. [3].
[2] Data from Iverson et al. [10].

Milk Fatty Acid Composition: Species Variability

Milk fat content and composition vary greatly among species and are probably best suited for the needs of each species. The milk of ruminants contains high levels of short chain fatty acids (SCFA), that of rodents and lagomorphs contains about one third MCFA while that of carnivores is practically devoid of fatty acids shorter than 14:0 (table 1). LCPUFA are derived from the diet, while MCFA and SCFA are synthesized in the mammary gland, the onset of synthesis being directly associated with parturition and independent of length of gestation in the human [5]. Different dietary input is evident from the differences between dog and ferret milk where the presence of fish products is associated with ω3 LCPUFA in ferret milk. Except for the absence of MCFA, there is remarkable similarity in fatty acid composition of carnivore (dog and ferret) and human milk, in spite of marked differences in milk fat content.

Distribution of fatty acids in triglycerides (which account for >98% of milk fat) affects the nature of the digestion products of gastric and pancreatic lipases, the former hydrolyzing preferentially the Sn-3 and the latter the Sn-1 and Sn-3 positions. Saturated LCFA are at the Sn-2 position, and would therefore be absorbed as monoglycerides in species whose milk does not contain digestive lipase which has no positional specificity and hydrolyzes with equal efficiency all three positions of triglyceride fatty acids.

Table 2. Extent of fat digestion in the stomach of the newborn

Species	Age	Diet	TG hydrolysis, %	Ref.
Pig	Newborn	Menhaden oil	45	9
Dog	2 weeks	Mother's milk	60.5	10
Seal	Newborn	Mother's milk	25–56	11
Human	Newborn [1]	Mother's milk	40	12

[1] Preterm infants 29.5 ± 0.8 weeks gestation, 4 weeks postnatal.

Gastric Lipolysis: Extent and Fatty Acid Specificity

Fat digestion starts in the stomach. A group of closely related enzymes originating either in the stomach (gastric lipase) or in oral tissues (lingual lipase and pregastric esterase) catalyzes the gastric hydrolysis of milk or formula fat in the newborn and of dietary fat in general in adults [6]. There is close homology (78%) among these enzymes and very little between them and pancreatic lipase [6]. Low pH optimum (2.5–6.5), absence of requirements for specific cofactors or bile salts and stability to pepsin, enable these enzymes to act in the stomach, and in certain diseases associated with pancreatic insufficiency (cystic fibrosis and chronic alcoholism [6–8]), also in the intestine. In vitro studies that simulate the gastric milieu have shown strong product inhibition, which limit lipolysis to 10–20% [6, 7]. In vivo studies that have actually measured the extent of gastric lipolysis show, however, that this process is extensive, resulting in hydrolysis of 40–60% of milk fat (table 2) [9–12]. Indeed, even the milk of aquatic mammals, with a fat content of 30–60% or menhaden oil, are digested to this extent in the stomach of the newborn.

Location of ω3 and ω6 LCPUFA at the Sn-3 position of milk triglyceride leads to their preferential release during gastric lipolysis [10, 11, 13]. Similar location of SCFA and MCFA leads also to their preferential release in the stomach, an observation that started the erroneous belief that gastric lipase and pregastric esterase are specific exclusively for MCFA and SCFA. The presence of only very low amounts of docosahexaenoic acid, (DHA, 22:6ω3) and very high levels of eicosapentaenoic acid (EPA, 20:5ω3) in free fatty acids (FFA) during gastric lipolysis in seals suggests retroconversion of DHA to EPA in the stomach [11]. Such retroconversion was previously reported in vitro as well as in vivo in animal and human studies, but the site of this process was not known [cited in 11]. The very high extent of gastric lipolysis (table 2) even in very premature infants [12] indicates that the adequate assimilation of fat as well as

Table 3. Gastric and milk-digestive lipase levels in several species

Lipase	Human	Dog	Ferret
Gastric, U/mg protein [1]	3.40 ± 0.96	4.25 ± 0.46	4.18 ± 0.57
	(1.89–4.77)	(1.80–8.32)	(1.80–9.20)
Milk, U/ml [2]	47 ± 8.0	30 ± 7.0	230 ± 10.0
	(24–116)	(20.3–41.3)	(180–240)

[1] Pinch biopsies of gastric fundus mucosa.
[2] Fresh milk. Lipolysis quantitated by release of ^{3}H-oleic acid from emulsified ^{3}H triolein, 1 U = 1 μmol fatty acid/min.

the adequate accretion of LCPUFA in milk-fed infants can be attributed in large measure to efficient gastric lipolysis.

Milk Digestive Lipase

A digestive lipase similar to pancreatic lipase was described in human and gorilla milk [14] and more recently in the milk of carnivores (dog, cat and ferret) [15, 16] (table 3). Because of a pH optimum of 7.5–9.0 and absolute bile salt dependence, the milk enzyme continues in the intestine the digestion of milk fat.

Although there are differences in molecular size between the human and carnivore milk enzymes, their characteristics are very similar. This milk digestive lipase is absent from the milk of many other species, such as rodents and ruminants [14, 15]. The enzyme is identical to pancreatic cholesterylester hydrolase, an enzyme extensively studied in several species.

Role of Gastric and Milk Lipases in the Digestion of Milk Fat

The fat in milk is contained within milk fat globules, the triglycerides (>98% of total milk fat) forming the core and the more polar lipids (phospholipid and cholesterol) and proteins the milk fat globule membrane. The latter is a barrier to pancreatic lipase, which cannot access the neutral fat within the globules [6]. We have reported earlier that the milk-digestive lipase, likewise, is unable to hydrolyze milk fat [17]. Both enzymes can, however, hydrolyze milk fat after initial lipolysis by lingual or gastric lipases, enzymes that can penetrate into the milk fat globule core [6].

Table 4. Initial hydrolysis (%) by gastric lipase is essential for subsequent hydrolysis of milk or formula fat by milk-digestive lipase

	Gastric lipase[1]	Milk-digestive lipase[2]	Gastric lipase and milk-digestive lipase[3]
Human milk	0.2–2.0	0.2–0.4	4.4–4.5
Dog milk	1.0–1.2	0.3–0.9	9.4–11.0
Ferret milk	2.3–3.5	0.4–1.4	34.0–48.0
Infant formulas			
S-26 (Wyeth-Ayerst)	1.5	0.3	10.8
Similac (Ross Labs)	1.2–2.0	0.3–0.7	5.7–7.7

[1] Incubation 15 min, pH 5.4.
[2] Incubation 15 min pH 8.5 + 12 m*M* taurocholate.
[3] Incubation 15 min pH 5.4, pH adjusted to 8.5, bile salt added and incubation continued for 15 min. Gastric lipase is inactive at pH 8.5, and milk digestive lipase at pH 5.4 [data from Kirk and Hamosh, FASEB J 1991;5:A1228].

The similarity of characteristics between carnivores (dog and ferret) and humans in milk fat composition (table 1), nature and activity level of preduodenal lipases (gastric) (table 2), and presence of milk-digestive lipase (table 3), enabled us to study whether, in general, predigestion by gastric lipase is a prerequisite for the action of milk lipase. The data show that even when milk lipase activity is very high (ferret), there is still an obligate need for initiation of this process by gastric lipase (table 4). The same interaction is necessary for the hydrolysis of formula fat, since the milk enzyme is unable to initiate the digestion of formula fat (table 4). The complementary characteristics of endogenous and exogenous (milk) lipases indicate that, as recently shown in vitro [18], milk fat can be completely digested, even when pancreatic lipase activity is low. Since heating does not impair the digestibility of milk fat or of isolated milk fat globules [19], recombinant milk digestive lipase could be added to pasteurized donor milk or to formula to improve fat digestion in premature infants.

References

1 Hamosh M, Bitman J, Liao TH, et al: Gastric lipolysis and fat absorption in preterm infants: Effect of MCT or LCT containing formulas. Pediatrics 1989;83:86–92.
2 Clandinin MT, Chappell JE, Heim T, et al: Fatty acid utilization in perinatal de novo synthesis of tissues. Early Hum Dev 1981;5:355–366.
3 Bitman J, Wood DL, Hamosh M, et al: Comparison of the lipid composition of breast milk from mothers of term and preterm infants. Am J Clin Nutr 1983;38:300–312.

4 Flores CA, Hing SAO, Wells MA, et al: Rates of triolein absorption in suckling and adult rats. Am J Physiol 1989;257:G823–G828.
5 Spear ML, Bitman J, Hamosh M, et al: Human mammary gland function at the onset of lactation: Medium chain fatty acid synthesis. Lipids 1992;27:908–911.
6 Hamosh M: Lingual and Gastric Lipases: Their Role in Fat Digestion. Boca Raton, CRC Press, 1990.
7 Hamosh M: Gastric and lingual lipases; in Johnson, Alpers, Jacobson, Christensen, Walsh (eds): Physiology of the Gastrointestinal Tract. New York, Raven Press, 1994.
8 Abrams CK, Hamosh M, Hubbard VS, et al: Lingual lipase in cystic fibrosis. Quantitation of enzyme activity in the upper small intestine of patients with exocrine pancreatic insufficiency. J Clin Invest 1984;73:374–382.
9 Chiang S-H, Pettigrew JE, Clarke SD, et al: Digestion and absorption of fish oil by neonatal piglets. J Nutr 1989;119:1741–1743.
10 Iverson SJ, Kirk CL, Hamosh M, et al: Milk lipid digestion in the neonatal dog: The combined actions of gastric and bile salt stimulated lipases. Biochim Biophys Acta 1991;1083:109–119.
11 Iverson SJ, Sampugna J, Oftedal OT: Positional specificity of gastric hydrolysis of long-chain n-3 polyunsaturated fatty acids of seal milk triglycerides. Lipids 1992;27:870–878.
12 Armand M, Hamosh M, Mehta NR, et al: Gastric lipolysis in premature infants, effects of diet: Human milk or formula. FASEB J 1993;7:A201.
13 Jensen RG, Clark RM, de Jong FA, et al: The lipolytic triad: Human lingual, breast milk and pancreatic lipase: physiological implications of their characteristics in digestion of dietary fats. J Pediatr Gastroenterol Nutr 1982;1:243–255.
14 Freudenberg E: Die Frauenmilch-Lipase. Bibl Paediatr. Basel, Karger, No 54, 1953.
15 Freed LM, York CM, Hamosh M, et al: Bile salt-stimulated lipase in non-primate milk: Longitudinal variation and lipase characteristics in cat and dog milk. Biochim Biopys Acta 1986; 878:209–215.
16 Ellis MA, Hamosh M: Bile salt stimulated lipase: Comparative studies in ferret milk and lactating mammary gland. Lipids 1992;27:917–922.
17 Hamosh P, Hamosh M: Differences in composition of preterm, term and weaning milk; in Xanthou M (ed): New Aspects of Nutrition in Infancy and Prematurity. Amsterdam, Elsevier, 1987, pp 129–141.
18 Bernback S, Blackberg L, Hernell O: The complete digestion of human milk triacylglycerol in vitro requires gastric lipase, pancreatic colipase-dependent lipase and bile salt stimulated lipase. J Clin Invest 1990;85:1221–1226.
19 Lough DS, Hamosh M, Philpott JR, et al: Are there changes in fat following heating of human milk that affect its digestion by the newborn? Is there a role for supplementation with recombinant lipases? Pediatr Res 1993;33:105A.

Dr. Margit Hamosh, Department of Pediatrics, Georgeton University Medical Center,
3800 Reservoir Road, NW, Washington, DC 20007 (USA)

Galli C, Simopoulos AP, Tremoli E (eds): Fatty Acids and Lipids: Biological Aspects.
World Rev Nutr Diet. Basel, Karger, 1994, vol 75, pp 92–95

The Ratio of Linoleic Acid to α-Linolenic Acid in Infant Formulas: Current Facts and Future Research Directions

Eric L. Lien

Wyeth-Ayerst Laboratories, Philadelphia, Pa., USA

The fatty acids of infant formula play two important roles in the rapidly growing neonate: they provide 50% of dietary energy and also provide essential fatty acids. Over 60 years ago, Burr and Burr [1] demonstrated the essentiality of the *ω6 fatty acids.* In recognition of the importance of ω6 fatty acids, numerous national and international regulatory agencies require specific levels of linoleic acid (LA) to be present in *infant formula* (in general, 300–500 mg/100 kcal). The essentiality of *ω3 fatty acids* has been more recently recognized, but increasing preclinical and clinical evidence indicates that this class of fatty acids is also important for the optimal development of the neonate [2]. More limited recommendations concerning α-linolenic acid (LNA) have been made; several groups have suggested that LA/LNA ratios of 4–10 [3] or 5–15 [4] are appropriate for infant formula.

Fatty Acid Profiles of Infant Formulas

All infant formulas meet the regulatory requirements for LA, but levels of LNA and the ratio of LA/LNA vary widely. Table 1 provides ratio data from a wide variety of countries, including those from Europe, North America, the Middle East and the Pacific Rim. Although many formulas conform to the above suggestions for LNA, it is clear that some selected formulas are substantially outside of this range. In fact, some formulas in these surveys have LA/LNA ratios of greater than 100:1.

Table 1. ω3 and ω6 fatty acid composition of formulas from various countries

	Weight percent		
	LA	LNA	ω6/ω3
Germany[a]	9.0–53.5	0.3–6.5	7.9–125.5
United States[b]	12.7–52.5	1.3–6.7	7.1–26.9
Malaysia[c]	13.1–32.6	0.2–4.4	7.4–65.5
Philippines[c]	9.5–27.6	0.2–1.7	9.6–101.9
Saudi Arabia[c]	12.1–32.4	0.2–4.1	8.0–161.9

[a] Koletzko and Bremer, Acta Pediatr Scand 1989;78:513–521.
[b] Data supplied by individual manufacturers (1992).
[c] Lien, Yuhas, Kuhlman, unpubl. data.

Table 2. Oils supplying essential fatty acids for infant formulas[a]

	Weight percent		
Fatty acid	soybean	corn	canola
C16:0	10.8	8.9	4.2
C18:0	3.9	1.6	2.0
C18:1	23.0	28.2	58.8
C18:2	55.3	58.4	20.7
C18:3	7.0	1.2	8.0
C20:1	–	–	1.5
C22:1	–	–	0.5

[a] Representative fatty acid profiles of vegetable oils. Profiles may vary depending on growing conditions and plant variety.

Essential Fatty Acid Oils

The levels of essential fatty acids in infant formulas are dependent on the oils used to provide such fatty acids, with soy and corn oil being the two most commonly employed oils (table 2). *Corn oil* is used in many powder preparations due to its oxidative stability [5]. Unfortunately, the level of LNA is low and the LA/LNA ratio is high in corn oil (table 2). *Soy oil* contains a more balanced proportion of LA to LNA, and if prepared under appropriate conditions can be used to produce a stable infant formula. *Canola oil* (low erucic acid rapeseed oil) is a possibility for use in infant formula. This oil is acceptable in

Europe, but is currently precluded from use in infant formula in the United States by the US Food and Drug Administration (FDA) [6]. Use of canola oil allows for greater flexibility in the LA/LNA ratio and also provides high levels of monounsaturated fatty acids. The data in table 2 are typical profiles of these oils. Depending on growing conditions and processing considerations, various ratios of LA/LNA for each of these oils may be obtained. For example, three infant formula fat blends, all containing corn oil as the primary essential fatty acid oil, had LA/LNA ratios of 39:1 [5], 48:1 [7] and 62:1 [8].

Effect of Essential Fatty Acids on Circulating and Tissue Fatty Acid Profiles

Circulating fatty acid profiles and tissue fatty acid accretion rates are determined by a complex interaction of variables. Dietary fatty acid content is of central importance in this process, especially in the case of essential fatty acids. Other variables, such as absorption efficiency, rate of fatty acid oxidation, and distribution of fatty acids for either storage or insertion into phospholipid membranes will also play a role in the eventual fate of fatty acids [9]. The use of corn oil versus soy oil as a source of essential fatty acids in infant formula has been addressed in several studies. Ponder et al. [5] observed only minor differences in circulating fatty acid levels when term infants were fed formulas containing these oils for 8 weeks. In contrast, Uauy et al. [7] found marked reductions in circulating docosahexaenoic acid (DHA) levels in preterm infants fed a corn oil-containing formula versus a soy oil-containing formula for approximately 5 weeks. We have examined this question further in rats receiving diets prepared with typical infant formula fat blends with essential fatty acids primarily provided by soy oil, corn oil, or a mixture [8]. The circulating ω3 long chain polyunsaturated fatty acids (LCPUFA) in the soy group were modestly higher than in the corn or corn/soy groups. More marked differences were noted in liver phosphatidylethanolamine (PE) and phosphatidylcholine (PC) DHA levels, suggesting that alterations in tissue accretion of fatty acids may be magnified in comparison to the circulating levels. In contrast, no differences in brain PE or PC fatty acid profiles were noted between groups. Due to the importance of eicosanoid signaling mechanisms in a variety of tissues and the complex interaction of ω3 LCPUFA with arachidonic acid (AA) in these pathways [10], care must be exercised in the generation of fat blends for infant formulas.

Conclusion

Essential fatty acids must be provided in all infant formulas. LA may be provided from a variety of oils, while LNA sources are more limited. Due to the possibility of generating fat blends with extremely high LA/LNA ratios, the use of corn oil in infant formulas should be discouraged. The optimal ratio of LA/LNA is an important continuing research question. The possibility of lowering this ratio to approximately 4:1 is the topic of current research [11] and requires further study. In addition, the inclusion of AA and DHA directly into formulas intended for both preterm and term infants is a continuing issue. As in the case of the C18 precursors, the levels and ratios of ω3 and ω6 LCPUFA remain to be optimized.

References

1 Burr GO, Burr MM: On the nature and role of the fatty acids essential in nutrition. J Biol Chem 1930;86:587–621.
2 Innis SM: Essential fatty acids in growth and development. Prog Lipid Res 1991;30:39–103.
3 Canada Health and Welfare, Health Protection Branch Bureau of Nutritional Sciences, Nutritional Recommendations, 1990, Ottawa.
4 Aggett PJ, Haschke F, Heine W, et al: Comment on the content and composition of lipids in infant formulas. Acta Paediatr Scand 1991;80:887–896.
5 Ponder DL, Innis SM, Benson JD: Docosahexaenoic acid status of term infants fed breast milk or infant formula containing soy oil or corn oil. Pediatr Res 1992;32:683–688.
6 Federal Register: January 28, 1985, vol 5, No 18, p 3753.
7 Uauy RD, Birch DG, Birch EE, et al: Effect of dietary omega-3 fatty acids on retinal function of very-low-birth-weight neonates. Pediatr Res 1990;28:485–492.
8 Lien EL, Yuhas R, Boyle F, et al: Effect of infant formula fat blends containing corn or soy oil on circulating and tissue fatty acid profiles in weanling rats (abstract). 1st Int Congr Int Soc Study of Fatty Acids and Lipids, Lugano 1993.
9 Innis SM, Plasma and red blood cell fatty acid values as indexes of essential fatty acids in the developing organs of infants fed with milk or formulas. J Pediatr 1992;120:S78–S86.
10 Kinsella JE, Lokesh B, Broughton S, et al: Dietary polyunsaturated fatty acids and eicosanoids: Potential effects on the modulation of inflammatory and immune cells: An overview. Nutrition 1990;6:24–44.
11 Clark KJ, Makrides M, Neumann MA, Gibson RA: Determination of the optimal ratio of linoleic acid to α-linolenic acid in infant formulas. J Pediatr 1992;120:S151–S158.

Eric L. Lien, PhD, Wyeth-Ayerst Laboratories, PO Box 8299, Philadelphia, PA 19101 (USA)

Galli C, Simopoulos AP, Tremoli E (eds): Fatty Acids and Lipids: Biological Aspects.
World Rev Nutr Diet. Basel, Karger, 1994, vol 75, pp 96–101

Infant Feeding: Antioxidant Aspects

Gayle Crozier

Nestec SA, Nestlé Research Centre, Lausanne, Switzerland

For reasons put forward and discussed in this volume, the importance of long chain ω3 polyunsaturated fatty acids (PUFA) in the development of the infant's brain and nervous tissue is now recognized, and the addition of these fatty acids to infant formula has been recently recommended [1]. By the nature of their methylene interrupted double bonds, PUFA are highly reactive in the presence of oxygen and catalysts, and the more unsaturated they are, the more reactive. Table 1 demonstrates the rapidity of the relative oxidation rate of 18 carbon fatty acids with increasing number of double bonds [2]. Docosahexaenoic (DHA) and eicosapentaenoic acids (EPA), with six and five double bonds respectively, are exceedingly rapidly oxidized [3]. This oxidative fragility presents some special challenges to the food industry.

Autooxidation

In most biological molecules the electrons of covalent bonds have opposite spins whereas the electrons of molecular O_2 have parallel spins. Oxygen can only accept electrons one at a time, and it is this so-called 'spin restriction' which permits only sequential addition of electrons and which leads to the formation of free radicals such as superoxide radical [4] in the presence of oxygen and an initiator present. An initiator, for example a trace element or photosensitizer, the spin restriction of O_2 can be overcome and reactive oxygen products are then formed. Trace elements are powerful initiators of lipid oxidation [5, 6] and certain of these are also essential nutrients for infants. Some such as iron and copper are present in infant formula in quantities quite capable of inducing autooxidation (table 2).

Table 1. Induction period and relative rate of oxidation of fatty acids containing different numbers of double bonds (fatty acids were kept at 25 °C) [2]

Fatty acid	Induction period, h	Oxidation rate (relative)
18:0	–	1
18:1 (9)	82	100
18:2 (9, 12)	19	1,200
18:3 (9, 12, 15)	1.34	2,500

Table 2. Concentration of metal which decreases the keeping time of lard by 50% at 98 °C [6]

Metal	ppm
Iron	0.6
Zinc	19.6
Copper	0.05
Manganese	0.6
Chromium	1.2
Nickel	2.2
Vanadium	3.0
Aluminum	5

The Nutritional Consequences of Autooxidation

Autooxidation can result in losses of the highly unsaturated fatty acids [7]. Because of their fragility, the measured losses of these fatty acids are even used as one of the most reliable indices of the oxidative status of any food containing them. In an infant formula, the oxidative losses of these bioactive fatty acids would obviously have a negative impact on nutritional value. However, the negative effects of oxidation are not limited to losses of these important fatty acids, since the resulting fat oxidation products can and do interact with other food components to the overall detriment of the product. When casein was incubated with linoleic acid ethyl ester for 4 days at 60 °C, losses of amino acids were extensive (table 3) [2]. Many of these amino acids are nutritionally

Table 3. Amino acid losses when casein was incubated with linoleic acid ethyl ester for 4 days at 60 °C [2]

Amino acid	% loss	Amino acid	% loss
Lys	50	Met	47
Ile	30	Phe	30
His	28	Thr	27
Ala	27	Gly	29
Tyr	27	Arg	29
Asp	29		

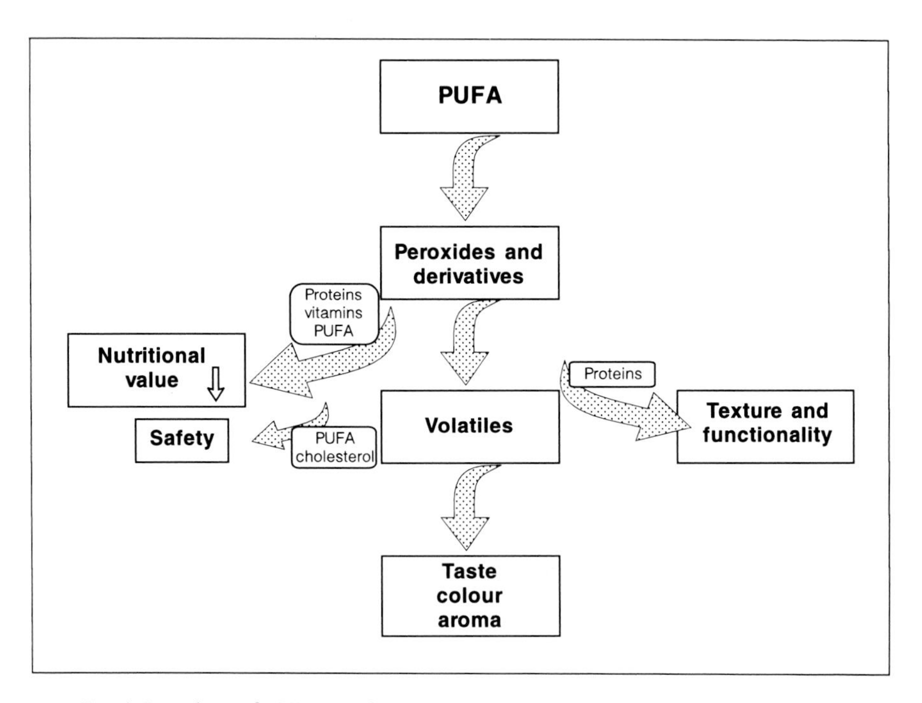

Fig. 1. Reactions of PUFA leading to nutritional and qualitative changes in foods.

essential amino acids, the losses of which deteriorate the nutritional value of this protein. Possible interactions involving proteins, vitamins, cholesterol as well as the PUFA are illustrated in figure 1. These oxidative reactions involving peroxides or derivatives of peroxides and resulting volatiles have been established to have negative impact on the taste, color, aroma, texture, functionality as well as the nutritional value and even safety of the food.

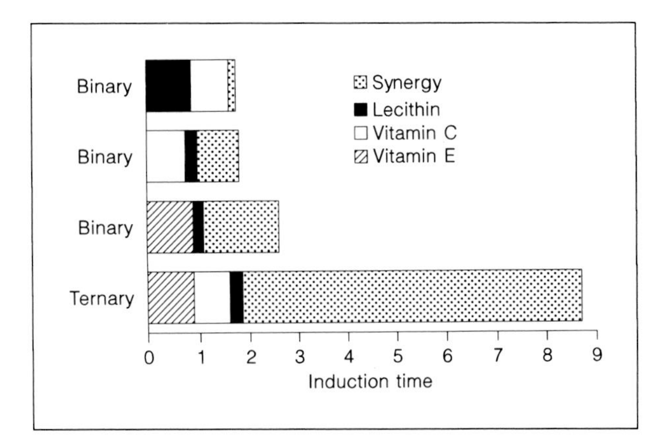

Fig. 2. Synergistic effects of antioxidant mixes in stabilizing marine oil. Marine oil was kept at 60 °C and oxidation was measured by the Fira-Astell oxygen absorption induction time test.

Antioxidants

Halliwell and Gutteridge [8] define an antioxidant as a substance that, when present at low concentrations compared to those of an oxidizable substrate, significantly delays or prevents oxidation of that substrate. This definition is rather good since it brings out the idea that the concept of an antioxidant is a relative one and each antioxidant must be considered in its context. For example, vitamin E, or α-tocopherol, is known to have good antioxidant activity in biological systems, but only weak activity in fats and oils [9]. Increasing the concentration of vitamin E in oils can paradoxically result in a prooxidant effect [10]. This can occur, for example, if α-tocopherol acts to reduce residual ferric ions to the more oxidatively reactive ferrous ion. It is thus important to choose an antioxidant which will be effective based on its mechanism of action and which is adapted to the final formulation. Due to legal reasons, the range of antioxidants potentially applicable to infant formula is restricted to the antioxidant vitamins. At Nestec we found and patented an effective antioxidant mixture of vitamin E, vitamin C and lecithin (fig .2). With any two of the components, the oxidative induction time of fish oil was between 2 and 3 h. A mixture of the three elements showed a remarkably effective synergism, roughly tripling the induction time. In this system vitamin C is used as a regenerator of the tocopheroxyl radicals formed by the primary antioxidant, vitamin E. Lecithin solubilizes the vitamin C in the lipid phase. With the addition to infant formula of long chain PUFA with increasing numbers of double bonds, the search for ever more effective antioxidant mixtures takes an even greater importance.

Table 4. Free ion in adults, term or preterm infants [14]

Group	n	Free plasma iron, μmol/l
Adult	18	0
Term infant	24	25
Preterm infant	22	48

The Antioxidant Status of the Infant

The premature infant is at a disadvantage compared to the term infant with respect to antioxidant protection. During the final 15% of gestation, in many different species, antioxidant enzymes increase markedly [11]. This appears also to be true in the human, as placental [12] and lung [13] superoxide dismutase activities reach their maximum at the end of pregnancy. Levels of antioxidant proteins such as transferrin [14] and ceruloplasmin [15] are also significantly diminished in preterm compared to term infants.

Plasma-free iron, a potential prooxidant, is much more elevated in the preterm than the term infant, and both are much higher than in the adult (table 4) [14], while levels of the antioxidant vitamins are low. Low vitamin E stores in the fetus are concomitant with its low level of adipose tissue stores [16] and requirements for vitamin C are greater in the preterm compared to the term infant [17]. The addition of the antioxidant vitamins to the PUFA-containing formula would protect not only the product but also the infant.

Conclusion

The challenge to the food industry is to understand the biochemical and physiological requirements of the infant and to use this knowledge to formulate appropriate products. One of the examples is the need of the infant for long chain PUFA for optimal growth and function of the brain and visual system. The sensitivity of these nutrients to oxidation represents a technological challenge which can be met by appropriate processing, handling of the product and the judicious choice of antioxidants.

The addition of the appropriate antioxidants is critical not only for the protection of these biologically active PUFA, but also to ensure the quality of the formula in terms of its overall nutritional value and organoleptic characteristics. These antioxidants also have potential for improving the antioxidant status of the infant.

References

1 British Nutrition Foundation: Unsaturated Fatty Acids. Nutritional and Physiological Significance. London, Chapman & Hall, 1992.
2 Beitz HD, Grosch W: Food Chemistry. Berlin, Springer, 1987.
3 Cho S-Y, Miyashita K, Miyazawa T, et al: Autooxidation of ethyl eicosapentaenoate and docosahexaenoate. JAOCS 1987;64:876–879.
4 Gutteridge JMC: Polyunsaturated Fatty Acids, Eicosanoids and Antioxidants in Biology and Human Disease. Proc IFSC Conf., Copenhagen, 1993.
5 Eriksson CE: Oxidation of lipids and food systems; in Chan HWS (ed); Autooxidation of Unsaturated Lipids. New York, Academic Press, 1987.
6 Hudson BF: Food Antioxidants. Amsterdam, Elsevier, 1990.
7 Han D, Ock-Sook Y, Shin HK: Solubilization of vitamin C in fish oil and synergistic effect with vitamin E in retarding oxidation. JAOCS 1991;68:740–743.
8 Halliwell B, Gutteridge JMC: How to characterize a biological antioxidant. Free Radic Res Commun 1990;9:1–32.
9 Pokorny J: Major factors affecting autoxidation; in Chan HW-S (ed): Autooxidation of Unsaturated Lipids. New York, Academic Press, 1987.
10 Yamamoto K, Niki E: Interaction of α-tocopherol with iron: Antioxidant and prooxidant effects of α-tocopherol in the oxidation of lipids in aqueous dispersions in the presence of iron. Biochim Biophys Acta 1988;958:19–23.
11 Frank L: Developmental aspects of experimental pulmonary oxygen toxicity. Free Radic Biol Med 1991;11:463–494.
12 Hien PV, Kovacs K, Matkovics B: Study of SOD activity changes in human placenta of different ages. Enzyme 1974;18:341–347.
13 McElroy M, Postle T, Kelly F: Antioxidant activity in fetal and neonatal lung; in Emerit I, Packer L, Auclair C (eds): Antioxidants in Therapy and Preventive Medicine. New York, Plenum Press, 1990, pp 449–454.
14 Moison RMW, Palinckx JJS, Roest M, et al: Induction of lipid peroxidation of pulmonary surfactant by plasma of preterm babies. Lancet 1993;341:79–82.
15 Hilderbrand DE, Fahim Z, James E, et al: Ceruloplasmin and alkaline phosphatase levels in cord serum of term, preterm and physiologically jaundiced neonates. Am J Obstet Gynecol 1974;118: 950–954.
16 Bell EF, Filer LJ: The role of vitamin E in the nutrition of premature infants. Am J Clin Nutr 1981; 34:414–422.
17 Arad ID, Ephraim S, Eyal FG: Plasma ascorbic acid levels in premature infants. Int J Vitam Nutr Res 1982;52:50–54.

Gayle Crozier, Nestec SA, Nestlé Research Centre, PO Box 44, Vers-chez-les-Blanc, CH–1000 Lausanne 26 (Switzerland)

Galli C, Simopoulos AP, Tremoli E (eds): Fatty Acids and Lipids: Biological Aspects.
World Rev Nutr Diet. Basel, Karger, 1994, vol 75, pp 102–104

The Influence of Maternal Vegetarian Diet on Essential Fatty Acid Status of the Newborn

Sheela Reddy, T.A.B. Sanders, Omar Obeid

Department of Nutrition and Dietetics, King's College London, UK

Low birth weight (<2,500 g) is more common in babies born to mothers of South Asian descent [1] compared with the United Kingdom (UK) general population, particularly within the subpopulation of Hindu women, the majority of whom follow a vegetarian diet. The duration of gestation amongst the Hindu women was approximately 4 days shorter compared with the indigenous UK population. We have previously reported lower proportions of docosahexaenoic acid (DHA, 22:6ω3) in plasma phospholipids and erythrocyte lipids of Caucasian strict vegetarians compared with omnivores [2]. It was suggested that this reflected the absence of DHA from the diet, competition from high intakes of linoleic acid (LA) and a limited capacity to synthesize DHA in man. The present study was carried out to see if the proportion of DHA was lower in the blood and tissues of life-long vegetarians and whether infants born to these vegetarians have lower proportions of DHA in their phospholipids.

Subjects and Methods

Cord blood samples and umbilical cords were collected from 146 deliveries (48 Asian vegetarians and 96 white omnivores) along with information on the pregnancy and delivery. Cord artery and plasma phospholipid fatty acid analyses were carried out on samples from 32 South Asian vegetarians and 32 nonvegetarian whites matched for age, parity, gender of newborn and gestational age. Multivariate regression analysis was carried out to search for any relationship between polyunsaturated fatty acids in cord plasma and artery phospholipids.

Table 1. Polyunsaturated fatty acids in cord artery and plasma phospholipids in Asian vegetarians and omnivores (mean values ± SEM)

Fatty acid	Cord plasma phospholipids, wt%		Cord artery phospholipids, wt%	
	vegetarians n = 27	omnivores n = 27	vegetarians n = 32	omnivores n = 32
20:3ω9	0.71 ± 0.102	0.60 ± 0.057	3.74 ± 0.187*	3.14 ± 0.182
20:4ω6	17.10 ± 0.80	15.71 ± 0.587	13.87 ± 0.354*	13.12 ± 0.329
22:5ω6	2.34 ± 0.158**	1.58 ± 0.126	4.15 ± 0.156**	3.19 ± 0.150
22:6ω3	4.01 ± 0.358**	5.84 ± 0.305	4.05 ± 0.173**	5.75 ± 0.191

*$p < 0.05$; **$p < 0.001$ compared with omnivores.

Results

Early onset of labour and emergency Caesarean section were significantly more common in the South Asian women, but otherwise both delivery characteristics were similar. The duration of gestation was 5.6 days shorter in the South Asian vegetarian women (unadjusted) as were birth weight and head circumference. Adjustments were made for confounding factors such as maternal age, gestational age and maternal height by analysis of variance. Between groups differences had a larger effect than smoking on birth weight and head circumference but a smaller effect on length. Mean birth weight, length and head circumference in the 32 matched pairs were 3,179 g, 51 and 34 cm in the South Asians, and 3,482 g, 53 and 34.7 cm in the white subjects. The proportion of docosapentaenoic acid (DPA, 22:5ω6) was significantly greater and that of DHA (22:6ω3) was lower in the cord artery and plasma phospholipids of the Asian vegetarians (table 1). The proportion of 20:3ω9 in plasma phospholipids was low in both groups but was elevated in cord artery phospholipids. Neither the proportion of DHA (22:6ω3) nor the ratio of 20:3ω9/20:4ω6 was significantly related to birth weight or head circumferences in multivariate analysis in the subset of 64 subjects.

Discussion

Birth weight, head circumference and length were all lower in the South Asian vegetarians even after correction for gestational age, gender of infant, parity, smoking habits, maternal age and height. Crawford et al. [3] have suggested that an inadequate intake of essential fatty acids might be related to low birth weight and that an elevated ratio of 20:3ω9/20:4ω6 in cord artery lipids was a sign of LA deficiency. However, we found this ratio to be elevated in all of the cord artery samples studied independent of birth weight. Yet this ratio was low in cord plasma phospholipids. Using multivariate analysis, we were unable to find any relationship between birth weight or length or head circumference and the proportion of DHA or the ratios of 20:3ω9/20:4ω6 and 22:5ω6/22:6ω3 in cord artery or cord plasma phospholipid in multivariate analyses. Therefore, we conclude the differences in maternal fatty acid intake are unlikely to explain the lower birth weight of the infants born to the vegetarian women in our study. The lower proportion of DHA was compensated for almost entirely by a higher proportion of DPA (22:5ω6), similar to observations made in animals fed diets with a high ratio of LA to α-linolenic acid. Our results demonstrate that vegetarians give birth to infants with less DHA in the blood and tissues. Whether these differences are of physiological significance or are within the normal physiological range is uncertain. However, these differences may need to be taken into account when formulating diets for the preterm infant.

Conclusion

Birth weight adjusted for gestational age is lower and the duration of gestation is shorter in Asian vegetarians than in white omnivores. Lower levels of DHA are found in the cord artery and plasma phospholipids of the vegetarians but are not related to birth weight in multivariate analysis.

References

1 McFadyen IR, Campbell-Brown M, Abraham R, et al: Factors affecting birthweight in Hindus, Moslems and Europeans. Br J Obstet Gynaecol 1984;91:968–972.

2 Sanders TAB, Ellis FR, Dickerson JWT: Studies of vegans: The fatty acid composition of plasma choline phosphoglycerides, erythrocytes, adipose tissue and breast milk and some indicators of susceptibility to ischaemic heart disease in vegans and controls. Am J Clin Nutr 1978;31:805–813.

3 Crawford MA, Doyle W, Drury P, et al: n-6 and n-3 fatty acids during early human development. J Intern Med 1989;225(suppl 1):159–169.

Dr. T.A.B. Sanders, Department of Nutrition and Dietetics, King's College London, Campden Hill Road, London W8 7AH (UK)

Galli C, Simopoulos AP, Tremoli E (eds): Fatty Acids and Lipids: Biological Aspects.
World Rev Nutr Diet. Basel, Karger, 1994, vol 75, pp 105–108

Mature Milk from Israeli Mothers Is Rich in Polyunsaturated Fatty Acids

Pierre Budowski[a], *Hannah Druckmann*[b], *Barry Kaplan*[c], *Paul Merlob*[b]

[a] Faculty of Agriculture, The Hebrew University of Jerusalem, Rehovot;
Departments of [b] Neonatology and [c] Obstetrics and Gynecology,
Beilinson Medical Center, Petah-Tiqva, and Sackler Faculty of Medicine,
Tel-Aviv University, Tel-Aviv, Israel

The present report presents data on the polyunsaturated fatty acid (PUFA) composition of mature milk in Israel which has a background of high linoleic acid (LA) consumption. Correlations between milk PUFA are also discussed.

Experimental

The donors were 26 healthy Jewish nursing women aged 23–43 years (mean 30.8). Gestation periods were 37–41 weeks and the newborns weighed over 1,500 g. Milk was sampled at the beginning and end of an evening feed, 6–10 weeks postpartum, and samples were methylated by the one-step acetyl chloride-methanol procedure [1]. Methyl esters were separated by programmed-temperature GLC on fused-silica capillary columns coated with Supelcowax 10. Student's t test was applied to differences between fore- and aftermilk values and correlations were obtained by the method of least squares [2].

Results

Table 1 shows the fatty acid (FA) profiles and total FA contents of fore- and hindmilk. The latter had twice the total FA content of foremilk, but the FA profile was the same. There was a close correlation between LA and α-linolenic acids (LNA) in milk (fig. 1). Other correlations are shown in table 2.

Table 1. FA profile of breast milk: FA concentrations are expressed as % of total FA (means ± SD)

	Foremilk n = 26	Aftermilk n = 26
Saturated FA		
C_8–C_{14}	16.49 ± 6.39	15.99 ± 5.15
16:0	20.16 ± 2.15	20.14 ± 2.42
18:0	5.81 ± 1.05	5.53 ± 1.09
$C_{20}+C_{22}$	0.23 ± 0.09	0.24 ± 0.08
Total saturated FA	42.70 ± 6.23	41.91 ± 5.42
Monounsaturated FA		
16:1	2.52 ± 0.64	2.63 ± 0.78
18:1	29.89 ± 3.09	30.07 ± 2.93
20:1	0.46 ± 0.15	0.47 ± 0.13
22:1	0.12 ± 0.15	0.10 ± 0.10
Total monounsaturated FA	32.99 ± 3.29	33.29 ± 3.25
Polyunsaturated FA		
18:2ω6	19.27 ± 3.87	20.02 ± 3.65
20:4ω6	0.60 ± 0.12	0.58 ± 0.12
Other LC ω6 FA	0.96 ± 0.25	1.03 ± 0.21
Total LC ω6 FA	1.57 ± 0.29	1.61 ± 0.26
Total ω6 FA	20.83 ± 3.99	21.63 ± 3.74
18:3ω3	1.50 ± 0.43	1.57 ± 0.44
22:6ω3	0.38 ± 0.16	0.37 ± 0.17
Other LC ω3 FA	0.37 ± 0.15	0.40 ± 0.17
Total LC ω3 FA	0.75 ± 0.24	0.77 ± 0.26
Total ω3 FA	2.26 ± 0.57	2.34 ± 0.54
Total LC PUFA	2.31 ± 0.56	2.39 ± 0.42
Total PUFA	23.09 ± 4.43	24.36 ± 4.44
PUFA ratios		
18:2ω6/18:3ω3 FA	13.30 ± 2.52	13.21 ± 2.37
LC ω6/LC ω3 FA	2.15 ± 0.68	2.28 ± 0.77
Total ω6/total ω3 FA	8.35 ± 1.61	9.47 ± 1.74
PUFA/saturated FA (P/S)	0.56 ± 0.17	0.59 ± 0.16
Total FA in milk, g/l	25.3 ± 13.6	51.5 ± 21.3*

* $p < 0.05$.

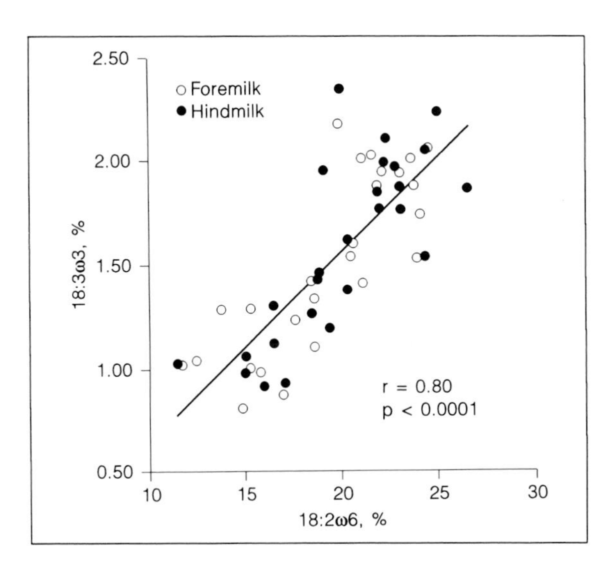

Fig. 1. Correlation between LA and LNA.

Table 2. Correlations between parent PUFA and LCPUFA

x	y	r	p
18:2ω6	20:4ω6	0.20	0.16
18:3ω3	22:6ω3	0.03	0.83
18:2ω6	Σ(LC ω6 FA)	0.43	0.002
18:3ω3	Σ(LC ω3 FA)	0.30	0.03
Σ(LC ω6 FA)	Σ(LC ω3 FA)	0.44	0.001

Discussion

The sampling procedure is adequate, as the FA composition of mature milk does not change during the feed (table 1), as already noted by others [3, 4], and also remains uniform during the day and during continuing lactation [3, 5]. The increase in total FA or milk fat (= total FA × 1.05) from fore- to aftermilk agrees with earlier work [4, 5].

The outstanding feature of milk in this study is the elevated LA level averaging 20% of the total FA, higher than the values recently compiled from studies of many other populations [6, 7]. LA in excess of 20% of total FA was however reported for milk from vegetarian and vegan women. LNA is also high, whereas arachidonic acid (AA) and docosahexaenoic acid (DHA) are the range of international values [6, 7]. The high PUFA content in turn results in an elevated P/S ratio, close to 0.6 (table 1). Of further interest is the marked uniformity of LA in our samples: the coefficient of variation (CV) calculated by Koletzko et al. [7] for LA in German breast milk is twice that of the present samples (35 vs. 18), while the CV values for AA are similar (19 and 20, respectively).

The fat consumption pattern in Israel is characterized by a high habitual LA intake, as best seen from the FA analysis of subcutaneous adipose tissue: four reports published since 1976 have yielded means ranging from 22 to 27% of total FA, with a weighted mean of 23.9% for 150 individuals. National food balance sheets [8] show that no separated animal fats are available and butter consumption is <0.5 en%, so that nearly two-thirds of the total food fat is of vegetable origin, including 12.5 en% vegetable oils. The most popular multipurpose fat is soybean oil, and other high LA oils are also used. We suggest that the high LA level in the milk, its low CV and its strong association with LNA are due to the above pattern.

The lack of correlation between the two parent PUFA and their main metabolites, AA 20:4ω6 and DHA 22:6ω3 (table 2), was noted by others; however, the association between LA and LNA (fig. 1) and that between the two parent PUFA and the sums of their respective long chain (LC) ω6 and LC ω3 metabolites (table 2) were not reported previously. They account, in part, for the correlation between Σ(LC ω6) and Σ(LC ω3) FA, noted by Koletzko et al. [7], and point to a regulation of the incorporation of both 20:4ω6 and DHA into milk lipids, whereas no such control seems to apply to the other LCPUFA.

Does the high milk LA content possibly interfere with the infant's metabolism of LC ω3 FA? Experience gained with low-birth-weight-infant formulas containing LA-rich vegetable oils supplemented with marine oils resulted in an adequate DHA status and satisfactory functional tests. Whether this is also the case with the milk studied here remains to be determined.

References

1 Lepage G, Roy CC: Improved recovery of fatty acids through direct transesterification without prior extraction or recovery. J Lipid Res 1984;25:1391–1396.
2 SAS User's Guide: Statistics, version ed 5. Cary, SAS Institute, 1985.
3 Hall B: Uniformity of human milk. Am J Clin Nutr 1979;32:304–312.
4 Koletzko B, Mrotzek M, Bremer HJ: Fat content and cis- and trans-isomeric fatty acids in human fore- and hindmilk; in Hamosh M, Goldman AM (eds): Human Lactation. II. Maternal and Environmental Factors. New York, Plenum, 1986, pp 589–594.
5 Harzer G, Haug M, Dieterich I, et al: Changing patterns of human milk lipids in the course of the lactation and during the day. Am J Clin Nutr 1983;37:612–621.
6 Innis SM: Human milk and formula fatty acids. J Pediatr 1992;120:S56–S61.
7 Koletzko B, Thiel I, Abiodun PO: The fatty acid composition of human milk in Europe and Africa. J Pediatr 1992;120:S62–S70.
8 Central Bureau of Statistics: Food Balance Sheets, Jerusalem, 1992.

P. Budowski, PhD, Faculty of Agriculture, PO Box 12, Rehovot 76–100 (Israel)

Galli C, Simopoulos AP, Tremoli E (eds): Fatty Acids and Lipids: Biological Aspects.
World Rev Nutr Diet. Basel, Karger, 1994, vol 75, pp 109–111

Molecular Species Composition of Plasma Phosphatidylcholine in Human Pregnancy

A.D. Postle[a], *G.C. Burdge*[a], *M.D.M. Al*[b]

[a] Child Health, University of Southampton, UK;
[b] Human Biology, University of Limburg, Maastricht, The Netherlands

Although the hyperlipidaemia of late pregnancy has been widely recognized for many years [1], the precise mechanisms and maternal adaptations involved in the direct transfer of polyunsaturated fatty acids (PUFA) from maternal liver to the fetus have been much less clearly defined. Fatty acids such as arachidonic (20:4) and docosahexaenoic (22:6) are enriched in phosphatidylcholine (PC) and it is possible that lipoprotein PC may act as a major carrier of PUFA in esterified form for their delivery from the liver to the peripheral tissues.

A role for selective placental transport has been proposed to contribute to the so-called 'biomagnification' of essential fatty acids between maternal and fetal plasma phospholipids [2], but the contribution of adaptations to the lipoprotein metabolism of the mother have not been studied in depth. Maternal liver and plasma PC composition in the pregnant rat show increased concentrations at term of molecular species containing *sn*-1 16:0 and *sn*-2 22:6 and 20:4 (PC16:0/22:6 and PC16:0/20:4), which precede the postnatal acquisition of these fatty acids in neonatal tissues [3]. Analysis of the individual species composition of phospholipids provides a distribution of the biologically and metabolically relevant phospholipid molecules.

In this study, we have extended previous measurements of plasma phospholipid fatty acids in human pregnancy [4] by analyzing the molecular species composition of maternal plasma PC throughout gestation.

Methods and Subjects

Serial blood samples were taken from 13 pregnant women, all of whom proceeded to normal term delivery without complication, at 15, 22 and 32 weeks of gestation, at delivery and 6–12 months postnatally. Corresponding samples of cord blood were also taken. A PC

Table 1. Concentration of unsaturated PC molecular species in maternal plasma during and after pregnancy (mean ± SD, n = 13)

Molecular species	Plasma PC, μ*M*				
	16 weeks	22 weeks	32 weeks	term	postnatal
16:0/22:6	140 ± 45	200 ± 50*	197 ± 37*	205 ± 54*	120 ± 46
16:0/20:4	197 ± 49	228 ± 50	214 ± 59	246 ± 73	197 ± 59
16:0/18:2	718 ± 182	868 ± 212*	924 ± 222*	984 ± 228*	764 ± 138
18:0/22:6	34 ± 18	41 ± 19	34 ± 19	24 ± 16	33 ± 21
18:0/20:4	115 ± 29	130 ± 27	109 ± 29	114 ± 33	130 ± 36
18:0/18:2	363 ± 94	404 ± 125	382 ± 122	357 ± 100	397 ± 80

*$p < 0.01$ vs. 16 weeks values by paired Students t test.

fraction was prepared by chloroform/methanol extraction, followed by selective adsorption to a BondElut NH_2 cartridge. PC molecular species were resolved by reversed phase HPLC and quantified by post-column derivatization with 1,6-diphenyl-1,3,5-hexatriene [5].

Results

Total plasma PC concentration increased steadily throughout pregnancy, from a value of 2.16 ± 0.40 m*M* at 16 weeks to a maximum of 2.93 ± 0.57 m*M* at delivery ($p < 0.001$), and declined postnatally (table 1). PC concentration in fetal cord plasma was considerably lower (1.28 ± 0.39 m*M*). The same 16 molecular species of PC were resolved in all plasma samples; over 50% of PC species in maternal plasma contained linoleate (18:2) at the *sn*-2 position, while fetal plasma was enriched in species containing 20:4 (31%). Concentrations in maternal plasma of the six PC species containing unsaturated fatty acid are given in table 1. The increase in plasma PC concentration was accounted for totally by a selective increase in species containing *sn*-1 16:0; concentrations of species containing *sn*-1 18:0 did not alter throughout gestation.

Discussion

The increase in *sn*-1 16:0 PC molecular species in maternal plasma in late gestation is consistent with a specific adaptation of maternal hepatic phospholipid metabolism. This change in plasma PC composition is coordinated with the acquisition of PUFAs into the phospholipids of fetal tissues. In the pregnant rat the molecular compositions of plasma and liver PC species are closely related, and are both characterized by increased *sn*-1 16:0 species, particularly 16:0/22:6, in late gestation [3]. Analysis of mechanisms of PC synthesis in rat [6] and guinea pig [7] liver show clearly that *sn*-1 16:0 species are the primary product of synthesis de novo from CDP:choline. These *sn*-1 16:0 PC species formed by synthesis de novo are converted to species containing 18:0 at the *sn*-1 position by the phospholipase-dependent process of acyl remodelling. Pulse:-chase substrate incorporation protocols suggest that such acyl remodelling requires at least 12 h to reach equilibrium [6]. Recent studies in our laboratory have demonstrated increased rates of PC synthesis in late gestation rat liver, coupled with an altered pattern of formation of selected PC species. For the rat, where PUFA supply is maximal in the immediate postnatal period, this adaptation of hepatic PC metabolism is probably caused both by the altered pattern of synthesis and a reduction in the extent of acyl remodelling. Increased flux through PC synthesis and an increased rate of hepatic lipoprotein synthesis and secretion may restrict the time available for PC acyl remodelling, and hence may lead directly to a redistribution from *sn*-1 18:0 to *sn*-1 16:0 PC species in late gestation. Extension of this argument to the human data presented above suggests that the similar selective increase of *sn*-1 16:0 PC species demonstrated in this study may be related to an increased synthesis and secretion of lipoprotein PC by maternal liver in late human gestation.

References

1 Darmady JM, Postle AD: Lipid metabolism in pregnancy. Br J Obstet Gynaecol 1982;89:211–215.
2 Crawford MA, Hassan AG, Williams G: Essential fatty acids and fetal brain growth. Lancet 1976;i: 452–453.
3 Hunt AN, Bellhouse AF, Kelly FJ, Postle AD: Late gestation changes in rat tissue phosphatidylcholine composition. Biochem Soc Trans 1991;19:111.
4 Al MDM, Hornstra G, van der Schouw YT, et al: Biochemical EFA status of mothers and their neonates after normal pregnancy. Early Hum Dev 1990;24:239–248.
5 Postle AD: Method for the sensitive analysis of individual molecular species of phosphatidylcholine by high performance liquid chromatography using post-column fluorescence detection. J Chromatogr 1987;415:241–251.
6 Tijburg LBM, Sambourski RW, Vance DE: Evidence that remodeling of the fatty acids of phosphatidylcholine is regulated in isolated rat hepatocytes and involves both the *sn*-1 and *sn*-2 position. Biochim Biophys Acta 1991;1085:184–190.
7 Burdge GC, Kelly FJ, Postle AD: Mechanisms of hepatic phosphatidylcholine synthesis in the developing guinea pig: Contributions of acyl remodelling and of N-methylation of phosphatidylethanolamine. Biochem J 1993;290:67–63.

Dr. A.D. Postle, PhD, Child Health, Southampton General Hospital,
Tremona Road, Southampton, Hants SO9 4XY (UK)

Galli C, Simopoulos AP, Tremoli E (eds): Fatty Acids and Lipids: Biological Aspects.
World Rev Nutr Diet. Basel, Karger, 1994, vol 75, pp 112–113

Summary Statement: Mechanisms of Accretion of Polyunsaturates in the Nervous System

R. E. Anderson

The session was co-chaired by *N. Salem,* Jr. and *R. E. Anderson,* and presentations were made by Drs. *Salem, Anderson, N. G. Bazan, S. A. Moore, E. Yavin,* and *F. Cockburn.*

The aim of the session was to explore the mechanisms by which neural tissues concentrate LCPUFA, especially those of the ω3 family. Discussions centered around the role and function of LCPUFA in these tissues. Key points made in the presentations were:

Humans can synthesize 20:4ω6 and 22:6ω3 from 18:2ω6 and 18:3ω3, respectively.

The retina and the brain have mechanisms for conservation of 22:6ω3 during ω3 deficiency. This is achieved in the eye by recycling 22:6ω3 between the retina and the retinal pigment epithelium.

The liver serves as a source of 22:6ω3 for the nervous system through synthesis from short chain precursors and incorporation into lipoproteins for delivery to target tissues. Alternatively, it was suggested that the liver may supply 22:5ω3 to the nervous system, especially when there is a limited intake of dietary 22:6ω3.

Tissues in the retina and the brain can synthesize 22:6ω3 from appropriate precursors. The relative contribution of local production and transport from the liver to the supply of LCPUFA to these tissues remains to be determined.

The levels of 22:6ω3 in brain and retina from infants who died of SIDS (sudden infant death syndrome) were significantly lower in those infants that had been raised on formula, compared to those raised on breast milk.

Recommendations from this symposium included the need for answers to the following questions:

What are the specific requirements for ω3 and ω6 PUFA by the developing human retina and brain? How are they supplied?

What are the long-term consequences of suboptimal accretion of 22:6ω3 in term and preterm human infants?

Are the biochemical changes of ω3 deficiency in human infants reversible?

What is the best form in which to supply ω3 and ω6 PUFA to those infants who are not breast fed?

Galli C, Simopoulos AP, Tremoli E (eds): Fatty Acids and Lipids: Biological Aspects.
World Rev Nutr Diet. Basel, Karger, 1994, vol 75, pp 114–119

Arachidonate and Docosahexaenoate Biosynthesis in Various Species and Compartments in vivo

Norman Salem, Jr., Robert J. Pawlosky

Laboratory of Membrane Biochemistry and Biophysics, Division of Intramural Clinical and Biological Research, National Institutes on Alcohol Abuse and Alcoholism, National Institutes of Health, Rockville, Md., USA

It has long been known that mammals are capable of elongating and desaturating 18-carbon essential fatty acids (EFA) such as linoleic (LA, 18:2ω6) and linolenic (LNA, 18:3ω3) acids to their 20- and 22-carbon end products. In fact, perhaps the first direct demonstration of desaturation in a living organism was made by Schoenheimer and Rittenberg [1] in 1936. They fed deuterated 18:0 to mice and subsequently demonstrated deuterium enrichment in an unsaturated fatty acid fraction prepared from the whole animal. Most subsequent work has been done using radioisotopically labeled 18:2ω6 or 18:3ω3 and has been largely confined to rats. In addition, most of this work has been performed in vitro. Emken et al. [2] have shown that stable isotopes can be used to measure 18:3ω3 metabolism in humans. In this lab, this approach has been improved with the use of a more sensitive mass spectral technique combined with a gas chromatographic system capable of separating the D_5-labeled metabolites from the otherwise identical 'unlabeled' endogenous molecules [3]. This approach facilitates the study of the in vivo metabolism of EFAs, particularly in large animals and humans where radioactive usage may be impractical or unsafe. In this communication, initial data are reported for a variety of mammalian species and it is demonstrated that all of these are capable of carrying out desaturation reactions in vivo.

Methods

Mice, rats, cats, rhesus monkeys and humans were placed on various diets where the fatty acyl composition was well defined. A single oral dose of deuterated 18:2ω6 (either [13,12,10,9-^{2}H]-18:2ω6 or [18,18,18,17,17-^{2}H]-18:2ω6) or deuterated 18:3ω3 ([18,18,18,17,17-^{2}H]-18:3ω3) were given with food. At various times, blood was collected (under ketamine anesthesia for rhesus) and subjected to lipid extraction by the method of Bligh and Dyer. In some cases, plasma was separated for lipid extraction. For rhesus monkeys and cats, liver needle biopsies were performed under ketamine anesthesia at various times. At sacrifice, brain, liver, retina, plasma and erythrocytes were excised and subjected to lipid extraction.

Lipid extracts were saponified, derivatized to the PFB ester and analyzed by GC/MS in the negative chemical ionization mode as described by Pawlosky et al. [3]. Data were collected using single ion monitoring at the expected retention time for each D_5-labeled metabolite and referenced to 23:0 as an internal standard. In this manner, subpicogram quantities of both the ω3 and ω6 metabolites could be simultaneously measured with very high selectivity. For the human studies, plasma samples were analyzed at 0, 8, 24, 48, 72, 96 and 168 after dosing with the deuterated fatty acids and the plot of the amount of each isotopically labeled fatty acid metabolite versus time was integrated; the values were then expressed as μg D_5 metabolite formed per ml of plasma.

Fatty acid composition was determined after methylation on an HP-5890 gas chromatograph with a flame ionization detector. Lipid extracts were evaporated with a stream of nitrogen and derivatized with BF_3 in methanol according to Morrison and Smith [4] except that hexane was added in place of benzene as cosolvent. The methyl esters were extracted into hexane and injected using a split capillary injection system at a 20:1 split ratio. Hydrogen was used as carrier gas at a linear velocity of 50 cm/s. The detector and injector temperatures were set at 250 °C. The oven ramp began at 140 °C for 2 min and then was ramped at 4 °C/min until 185 °C was reached. After 2 min, the temperature was ramped at 1 °C/min until 215 °C and finally increased at a rate of 30 °C/min until 245 °C. A DB-FFAP fused silica capillary column was used of 30 m×0.25 mm OD with a film thickness of 25 μm (J&W Scientific). Retention times of peaks were compared to authentic standards prepared by Nu Chek Prep (Elysian, Minn, USA) or analyzed with an HP-5970 mass selective detector in the electron impact mode.

Results

Mice, rats, cats, rhesus monkeys and humans are all capable of elongating, desaturating and transporting to the bloodstream 18:2ω6 and 18:3ω3 (table 1). It is well known that rodents have this activity [5] but it has been suggested that cats lack desaturase enzymes. Our data show that rats and mice exhibit substantial EFA metabolic activity even when they are on a chow diet that contains LC-PUFAs like 20:4ω6 and 22:6ω3. They are capable of all three sequential desaturation steps as both plasma, liver and brain 22:5ω6 and 22:6ω3 incorporate deuterium from their respective 18-carbon precursors. It was also clear that deuterated 18:2ω6 and 18:3ω3 are both taken up into brain after a single oral dose in young rats. Perfusion and subcellular fractionation studies show that these fatty acids are indeed in the brain parenchyma and are

Table 1. Formation of plasma/blood LC-PUFAs from deuterium-labeled 18-carbon precursors in various mammals

	Fatty acid metabolite, µg/ml						
	mouse blood[a]	rat blood[b]	feline plasma[c]		rhesus plasma[d]		human plasma[e]
Diet	chow	chow	chow	1% corn oil	chow	4% olive oil	ad libitum
Metabolite formed							
D_5-20:4ω6	0.88	0.28	ND[f]	0.04	0.19	0.3	1.9
D_5-20:5ω3	1.8	0.68	ND	1.95	0.7	1.88	3.8
D_5-22:6ω3	3.5	1.1	ND	ND	ND	0.4	0.3

[a] Expressed as the total amount in blood 48 h after an oral dose of 10 mg each of D_5-18:2ω6 and D_5-18:3ω3 ethyl ester, mean of n = 5 animals.
[b] Expressed as the total amount in blood 48 h after an oral dose of 10 mg each of D_5-18:2ω6 and D_5-18:3ω3 ethyl ester; mean of n = 4 animals.
[c] Expressed as the total amount in plasma 48 h after an oral dose of 100 mg each of D_5-18:2ω6 and D_5-18:3ω3 ethyl ester; mean of n = 4 animals. The 'chow' fed animals received a chow containing LC-PUFA, the experimental diet contained 1 wt% corn oil and 9 wt% hydrogenated coconut oil as the fat sources.
[d] Expressed as the total amount in plasma 48 h after an oral dose of 100 mg each of D_5-18:2ω6 and D_5-18:3ω3 ethyl ester; mean of n = 6 animals. The 'chow' fed animals received the NIH Primate Diet containing LC-PUFA; the experimental diet contained 4 wt% olive oil and 2 wt% hydrogenated coconut oil as the fat sources.
[e] Expressed as the area under the time course curve over the first 7 days after an oral dose of 1 g each of D_5-18:2ω6 and D_5-18:3ω3 ethyl ester; mean of n = 8 normal volunteers on an ad libitum American diet.
[f] ND indicates not detected.

not confined to cerebral blood vessels. A study of the ω3 metabolites incorporating deuterium versus time indicated a movement of the label through the pathway with mainly 22:6ω3 labeling at the end of 7 days. These observations indicated that the pathway in which 18:3ω3 is taken up and metabolized in brain to 22:6ω3 is operative, as is the analogous ω6 pathway.

Contrary to previous reports [6], domestic cats do express desaturase activity [7]. They are capable of forming 20:4ω6 and 22:5ω3 from 18:2ω6 and 18:3ω3, respectively. Very little activity could be detected when the animals were fed a chow diet containing 20:4ω6 and 22:6ω3 (table 1). However, when they were placed on an EFA-limited diet containing 1 wt% corn oil as the only fat source, significant induction of EFA metabolism occurred. It was of great interest that the 'Δ4' desaturase step did not take place even under the best case scenario, i.e., LC-PUFA elimination and low EFA. It is concluded that cat liver lacks Δ4-desaturase or as postulated by Voss et al. [8], the 24-carbon Δ6-desaturase [8]. It is also of importance that deuterated 22:6ω3, 22:5ω6 and their 24-carbon pentaene and hexaene precursors were detected in the brains of these cats. This was interpreted as indicating that the final step of desaturation to form 22:6ω3 can take place only in the cat nervous system. It also lends strong support to the mechanism for 22:6ω3 and 22:5ω6 biosynthesis proposed by Voss et al. [8] since the 24-carbon LC-PUFAs were not found in the plasma or liver but were found in the nervous system where the end products were formed.

In rhesus monkeys, both 20:4ω6 and 22:6ω3 were found to incorporate deuterium from their respective 18-carbon precursors in both plasma and liver when on a low EFA diet; in this case, the EFA was supplied by 4 wt% olive oil. When fed a primate chow diet containing LC-PUFA, no deuterium incorporation into 22:6ω3 nor 22:5ω6 was found in liver biopsy samples or plasma. Again it appeared that LC-PUFA in the diet led to abolition of their endogenous biosynthesis and accumulation.

Little in this regard has been published for humans but Emken et al. [2] have suggested that there is little or no conversion of 18:2ω6 to 20:4ω6 although 18:3ω3 conversion to 20:5ω3 and ultimately 22:6ω3 was observed [2]. In our human experiments, incorporation of deuterium into plasma 20:4ω6 is clearly observed after a single 1 g dose of D_5-18:2ω6. It was observed under all three of our dietary conditions, i.e., an ad libitum diet, a beef-based diet and a fish-based diet (data not shown). One observation that helps to reconcile our results with those of Emken et al. [2] is that the maximal plasma 20:4ω6 incorporation of deuterium from a single dose of 18:2ω6 is at 72–96 h. Since blood samples were only obtained for the first 48 h in Emken's study, the peak formation was missed. However, this by itself would not be an adequate explanation since a significant incorporation was also observed at shorter times in our study. It is believed that the use of a precursor with five deuterium atoms is the principal reason for this difference since it leads to a much higher signal-to-noise ratio as it can be completely resolved from the 'unlabeled' endogenous compound on a capillary column. This difference in retention, coupled with the much higher sensitivity (at least three orders of magnitude) of the negative chemical ionization method and the decrease in mass overlap from the natural abundance (in comparison to the D_2 or D_4 compounds used by Emken et al. [2]) leads to a much easier detection of small isotopically labeled peaks.

It should be pointed out that a very small peak can represent substantial flux through a metabolic pathway since there is a very high isotopic dilution with endogenous molecules. This dilution is at least an order of magnitude higher for 18:2ω6 in comparison to 18:3ω3 and so fewer deuterium-labeled molecules will be converted for the ω6 pathway for the same amount of activity with respect to the ω3 pathway. The fact that there is only about half as much D_5-labeled ω6 metabolite as for the ω3 pathway does not therefore suggest that the ω3 pathway is twice as active. On the contrary, it suggests that the ω6 pathway may be more active since there appears to be a more than twofold difference in isotopic dilution. This analysis cannot be adequately made without knowing the size of the relevant substrate pool and the whole body amount of product. Thus, the present analysis of plasma metabolite levels can be used mainly for qualitative comparisons.

Conclusions

In a broad survey of mammals from mice to man, it was demonstrated that all contain the capacity to elaborate EFAs in vivo. Cats do have elongation and desaturase activity at a lower level than most other mammals. Diet is a crucial modulating factor as LC-PUFA in the diet generally has the effect of decreasing 18-carbon EFA elaboration. Humans clearly possess both ω3 and ω6 EFA metabolic activity as deuterated 20:4ω6 and 22:6ω3 appear in the plasma.

Finally, the principal issue of this symposium should be directly addressed, i.e., the mechanisms involved in the accumulation of 22:6ω3 into brain. There is perhaps no direct data that allows us to understand the quantitative importance of the contributions of various pathways towards the accumulation of neural 22:6ω3. Different balances of direct dietary incorporation or maternal transfer of preformed 22:6ω3, liver partial or complete ω3 metabolism or brain partial or complete metabolism of 18:3ω3 or other ω3 intermediates may be operative. This balance will obviously be affected by the diet and by the maternal diet when the fetus is concerned and by whether human or artificial milk is fed. It may be expected that accretion of 22:6ω3 during early development may involve a different mix of these pathways than does its maintenance in the mature mammalian brain.

The example given by cats in which the ω3 and ω6 metabolic endpoints in the plasma are 22:5ω3 and 22:4ω6, respectively, is an extreme case but one which is suggestive of an alternative pathway for neural 22:6ω3 accumulation that may be applicable to all mammals that has not been considered before. The suggestion is that when there is insufficient preformed 22:6ω3 available in the diet, one pathway that is of quantitative importance is the biosynthesis of 22:5ω3 by the liver. This fatty acid is then exported through the plasma into the brain or retina where the final sequence of desaturation reactions occur. The brain 22:6ω3 level is the result of the synergistic metabolism of ω3 intermediates by the liver and brain. The nervous system may be specialized for the efficient functioning of this last metabolic step for 22:6ω3 and 22:5ω6 formation. Genetic or nutritional disturbances of this system may lead to neuropathology.

References

1 Schoenheimer R, Rittenberg D: Deuterium as an indicator in the study of intermediary metabolism. J Biol Chem 1936;113:505–510.
2 Emken EA, Adlof RO, Rakoff H, Rohwedder WK: Metabolism of deuterium-labeled linolenic, linoleic, oleic, stearic and palmitic acids in human subjects; in Baillie TA, Jones JR (eds): Synthesis and Applications of Isotopically Labeled Compounds 1988: Proceedings of the Third International Symposium. Amsterdam, Elsevier, 1988, pp 713–716.
3 Pawlosky RJ, Sprecher HW, Salem N, Jr: High sensitivity negative ion GC-MS method for detection of desaturated and chain-elongated products of deuterated linoleic and linolenic acids. J Lipid Res 1992;33:1711–1717.
4 Morrison WR, Smith LM: Preparation of fatty acid methyl esters and dimethylacetals from lipids with boron fluoride-methanol. J. Lipid Res 1964;5:600–608.
5 Naughton JM: Supply of polyenoic fatty acids to the brain: The ease of conversion of the short-chain essential fatty acids to their longer chain polyunsaturated metabolites in liver, brain, placenta and blood. Int J Biochem 1981;13:21–32.
6 Rivers JPW, Sinclair AJ, Crawford MA: Inability of the cat to desaturate essential fatty acids. Nature 1975;258:171–173.
7 Pawlosky RJ, Salem N, Jr: Metabolism of essential fatty acids in mammals; in Sinclair A, Gibson R (eds): Essential Fatty Acids and Eicosanoids. Champaign, American Oil Chemists' Society, 1992, pp 26–30.
8 Voss A, Reinhart M, Sankarappa S, Sprecher HW: The metabolism of 7,10,13,16,19-docosapentaenoic acid to 4,7,10,13,16,19-docosahexaenoic acid in rat liver is independent of Δ4-desaturase. J Biol Chem 1991;266:19995–20000.

Norman Salem, Jr., PhD, LMBB, NIAAA, Room 55C, 12501 Washington Avenue,
Rockville, MD 20852 (USA)

Galli C, Simopoulos AP, Tremoli E (eds): Fatty Acids and Lipids: Biological Aspects.
World Rev Nutr Diet. Basel, Karger, 1994, vol 75, pp 120–123

Docosahexaenoic Acid Supply to the Retina and Its Conservation in Photoreceptor Cells by Active Retinal Pigment Epithelium-Mediated Recycling

Nicolas G. Bazan, Elena B. Rodriguez de Turco, William C. Gordon

LSU Eye Center and Neuroscience Center, New Orleans, La., USA

Membranes of photoreceptor cells, highly enriched in docosahexaenoic acid (DHA) phospholipids [1], are being renewed each day as new membranes are added to the base of outer segments and photoreceptor tips are shed and phagocytized by the retinal pigment epithelium (RPE). Adequate DHA availability for photoreceptors is assured by: (a) active synthesis of DHA by the liver from dietary precursors (i.e. α-linolenic acid [2]); (b) efficient delivery of DHA incorporated into circulating plasma lipoproteins [2, 3] (referred to as the *long loop* [4]); (c) selective uptake by RPE [5, 6]; and (d) active recycling from RPE to retina (*short loop* [7]) after the daily shedding event [8, 9]. DHA trafficking from liver to retina and brain is very active in developing postnatal mice [2, Martin et al., submitted], and efficiently supports the large requirements for this essential fatty acid as new photoreceptor membranes are assembled [3].

Trafficking of DHA through the *Long* and *Short Loops* in the Frog

Frogs *(Rana pipiens)*, injected in the dorsal lymph sac with [^{3}H]DHA (50 μCi/g b.w.), were used to study DHA delivery to the retina for daily photoreceptor membrane renewal. In the frog, liver labeling peaked at 5 days, with a subsequent decrease to 50% by day 14. A similar early peak in liver [^{3}H]DHA labeling was reported for mouse pups [2] and young adult rats [10]. Total labeling of plasma lipids followed a profile similar to that of the liver, with a tissue half-life of 15 days. These systemic changes in labeled [^{3}H]DHA lipids

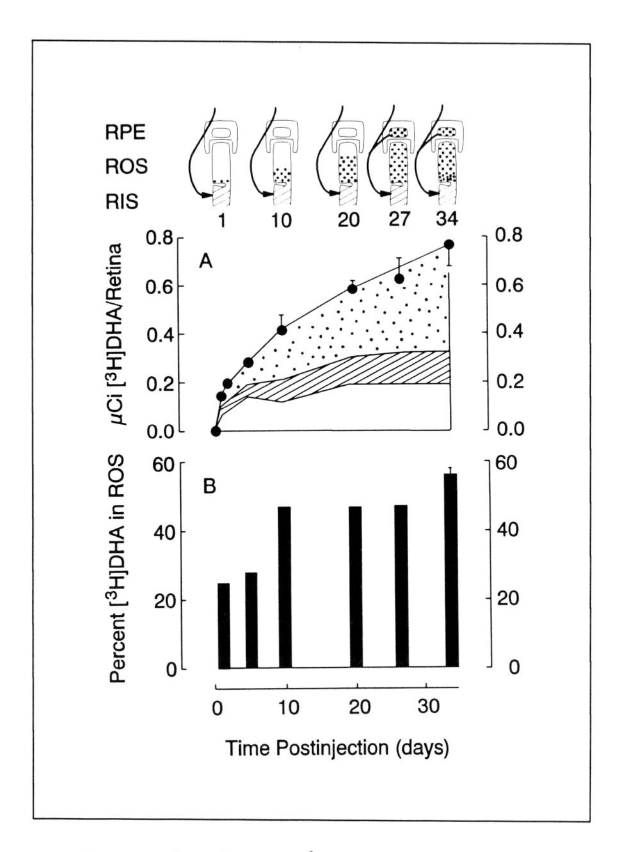

Fig 1. Profile of retinal [^{3}H]DHA labeling as a function of time. Frogs *(R. pipiens)* were injected with [^{3}H]DHA (50 μCi/g b.w.) in the dorsal lymph sac and retinal labeling was analyzed at different times postinjection. Diagrams of DHA trafficking through RPE cells via the *long loop* (from liver to photoreceptors) and *short loop* (from phagosomal lipids to photoreceptors). Densely labeled phagosomes (dotted) appear in RPE daily after 27 days postinjection. *A* Distribution of [^{3}H]DHA for 34 days postinjection, based on biochemical and autoradiographic analysis. ● = Total [^{3}H]DHA recovered per retina; ▨ = distribution of labeling in ROS; ▨ = rest of photoreceptors; □ = neural retina. *B* Percent labeling of ROS, with respect to total retinal labeling, determined by autoradiography.

occurred in parallel with gradual retinal uptake of the precursor, and were followed for 34 days after injection (fig. 1). Autoradiographic and biochemical analyses revealed a dramatic, rapid uptake of [^{3}H]DHA by the RPE, followed by a time-dependent, preferential labeling of rod outer segments (ROS).

There is continual assembly of new disc membranes at the base of outer segments. These become highly labeled with [^{3}H]DHA phospholipids from the constant supply of DHA from the liver. In fact, a densely labeled area grows apically until it fills the ROS [11, 12]. Between 10 and 27–28 days, when these heavily labeled discs reached the ROS tips, 46% of total labeling was detected autoradiographically in ROS (fig. 1). During the subsequent 7 days, densely labeled phagosomes appeared in RPE after each shedding event [11, 13], while the retina continued to accumulate label preferentially in ROS (56% of total [^{3}H]DHA label at day 34). Recently, very-long-exposure autoradiograms have revealed a second wave of dense labeling at the ROS base [unpubl. observation]. This is the first direct cytological evidence for *short loop* involvement in the recycling of DHA from densely labeled phagosomes back to photoreceptors.

Conclusions

This study shows that liver has the ability to accumulate DHA and gradually release it as part of circulating plasma lipoproteins. RPE is endowed with the ability to selectively and efficiently take up DHA from plasma and deliver it to photoreceptors through the interphotoreceptor matrix. This *long loop* is closed when DHA, released into circulation by RPE, is recovered by the liver. The *short loop*, which involves the recycling of phagosomal DHA lipids back to photoreceptors, will lead to conservation of this essential fatty acid within the retina. These two routes, i.e. the *short* and *long loops*, work in concert to supply sufficient DHA to photoreceptors for lipid synthesis and membrane biogenesis.

References

1 Aveldaño de Caldironi MI, Bazan NG: Composition and biosynthesis of molecular species of retina phosphoglycerides. Neurochemistry 1980;1:381–392.
2 Scott BL, Bazan NG: Membrane docosahexaenoate is supplied to the developing brain and retina by the liver. Proc Natl Acad Sci USA 1989;86:2903–2907.
3 Bazan NG, Gordon WC, Rodriguez de Turco EB: Docosahexaenoic acid uptake and metabolism in photoreceptors: Retinal conservation by an efficient RPE cell-mediated recycling process; in Bazan NG, Marfi MG, Toffano G (eds): Neurobiology of Essential Fatty Acids. Adv Exp Med Biol. New York, Plenum, 1992, vol 318, pp 295–306.
4 Bazan NG: The identification of a new biochemical alteration early in the differentiation of visual cells in inherited retinal degeneration; in LaVail MM, Anderson RE, Hollyfield JG (eds): Inherited and Environmentally Induced Retinal Degenerations. New York, Liss, 1989, pp 191–215.
5 Wang N, Anderson RE: Enrichment of polyunsaturated fatty acids from rat retinal pigment epithelium to rod outer segments. Current Eye Res 1993;11:783–791.
6 Wang N, Wiegand RD, Anderson RE: Uptake of 22-carbon fatty acids into rat retina and brain. Exp Eye Res 1992;54:933–939.
7 Bazan NG, Birkle DL, Reddy TS: Biochemical and nutritional aspects of the metabolism of polyunsaturated fatty acids and phospholipids in experimental models of retinal degeneration; in LaVail MM, Anderson RE, Hollyfield JG (eds): Retinal Degeneration: Experimental and Clinical Studies. New York, Liss, 1985, pp 159–187.

8 Gordon WC, Rodriguez de Turco EB, Bazan NG: Retinal pigment epithelial cells play a central role in the conservation of docosahexaenoic acid by photoreceptor cells after shedding and phagocytosis. Current Eye Res 1992;11:73–83.
9 Stinson AM, Wiegand RD, Anderson RE: Recycling of docosahexaenoic acid in rat retinas during n-3 fatty acid deficiency. J Lipid Res 1991;32:2009–2017.
10 Li J, Wetzel MG, O'Brien PJ: Transport of n-3 fatty acids from the intestine to the retina in rats. J Lipid Res 1992;33:539–548.
11 Gordon WC, Bazan NG: Docosahexaenoic acid utilization during rod photoreceptor cell renewal. J Neurosci 1990;10:2190–2202.
12 Gordon WC, Bazan NG: [^{3}H]docosahexaenoic acid uptake and utilization by retinal pigment epithelium and photoreceptors. Invest Ophthalmol Vis Sci 1993;34:2402–2411.
13 Bazan NG, Rodriguez de Turco EB, Gordon WC: Retinal pigment epithelium contributes to the conservation of docosahexaenoic acid by photoreceptor cells. Invest Ophthalmol Vis Sci 1992;33 (ARVO suppl):914.

Nicolas G. Bazan, MD, PhD, LSU Eye Center and Neuroscience Center,
2020 Gravier Street, Suite B, New Orleans, LA 70112 (USA)

Galli C, Simopoulos AP, Tremoli E (eds): Fatty Acids and Lipids: Biological Aspects.
World Rev Nutr Diet. Basel, Karger, 1994, vol 75, pp 124–127

The Accretion of Docosahexaenoic Acid in the Retina

Robert E. Anderson[a,b], *Huiming Chen*[a], *Nan Wang*[a], *Ann Stinson*[a]

[a] Department of Biochemistry and [b] Cullen Eye Institute, Baylor College of Medicine, Houston, Tex., USA

The retina contains the highest levels of docosahexaenoic acid (DHA, 22:6ω3) of any organ in the body, and the largest amounts are found in the outer segments (ROS) of rod photoreceptor cells [1]. These organelles contain the visual pigment rhodopsin and are responsible for transducing photons of light into electrical currents which are transmitted to the brain. Some years ago, we demonstrated that retinal function, measured by electroretinography, was dependent upon a dietary source of ω3 polyunsaturated fatty acids (PUFAs) [2]. Several other laboratories have demonstrated similar phenomena in rats [3], monkeys [4], and human infants [5]. Given an important, albeit unknown, function of DHA in the retina, it is not surprising that this tissue can concentrate and conserve DHA during ω3 deprivation [6]. The ability of the retina to concentrate DHA is shown in figure 1. In the frog, the blood contains low levels of DHA and arachidonic acid (AA, 20:4ω6). Both PUFAs are concentrated in the retinal pigment epithelium (RPE), a tissue involved in transport of nutrients from the blood to the retina. However, DHA is further concentrated in the retina and ROS, whereas AA is not. Interestingly, the only time ω6 PUFAs accumulate in the retina is during prolonged ω3 deficiency, when 22:5ω6 replaces 22:6ω3. This phenomenon has been observed in the brain by Galli et al. [7].

Conservation of DHA in the Retina

ROS are dynamic structures whose components are renewed daily [1]. Distal tips of ROS are shed and phagocytized by the RPE and newly synthe-

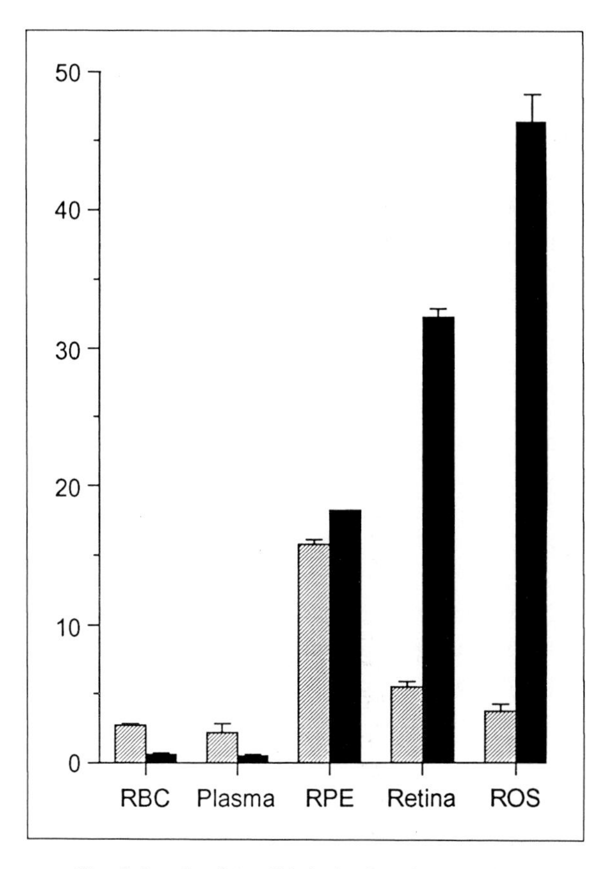

Fig. 1. Levels of AA (20:4ω6; ▨) and DHA (22:6ω3; ■) in total lipids of frog red blood cells (RBC), plasma, RPE, retina, and ROS. Values are means ± SD of three independent determinations.

sized membranes are added at the base of the ROS to replace those lost each day. This process is not affected by PUFA status. Recent studies have suggested that DHA is conserved in the retina through recycling between the retina and the RPE. We found that the DHA content of RPE increased significantly after a light-induced shedding event, but returned to baseline values within 8 h [8]. Bazan et al. [9] used autoradiography to show that labeled DHA was present in ROS tips that had been phagocytized by the RPE. In the RPE, ROS lipids are rapidly hydrolyzed and the released fatty acids incorporated into glycerolipids [Chen and Anderson, unpubl. results]. We compared the fates of DHA and AA and found that DHA was incorporated preferentially into triglycerides (TG), whereas AA went mainly into phospholipids. Also, DHA was incorporated through de novo synthesis, while AA was metabolized via deacylation-reacylation pathways. Studies with labeled glycerol revealed that over 50% of the label was incorporated into TG in the RPE, more than would be predicted from the steady-state level of TG (10%). We believe that the oil droplets containing TG provide efficient, short-term storage for DHA, which is shuttled back to the retina before the next shedding event. The differential metabolism of DHA and AA in the RPE may explain in part the enrichment of DHA in the retina relative to AA. However, the mechanism of maintaining the tenfold difference in the levels of DHA and AA in the ROS remains a mystery.

Sources of Retinal DHA

DHA can be supplied through the diet or by synthesis from ω3 precursors. The liver makes DHA from shorter chain ω3 fatty acids and has been suggested to be the major source of DHA for the retina and brain [9]. However, it is now clear that tissues of both the eye and the brain can synthesize DHA, if the appropriate precursors are available. In the brain, blood vessel endothelial cells can convert 18:3ω3 to 22:5ω3, but cannot accomplish the final desaturation step, which is carried out by astrocytes [10]. Neurons are also capable of performing some of the elongation-desaturation reactions, but are unable to produce DHA [10]. We have found that the RPE, but not the retina, can synthesize DHA from 18:3ω3 and 22:5ω3 [Wang and Anderson, unpubl. results]. The significance of astrocytes and RPE as sources of DHA must in part depend on the availability of appropriate precursors to these cells. Since the RPE has measurable amounts of 18:3ω3, 20:5ω3, and 22:5ω3, it seems likely that substantial amounts of DHA are synthesized and supplied by the RPE to the retina.

Conclusions

DHA is an important component of rod photoreceptor membranes. During ω3 deprivation, the eye conserves DHA by recycling between the retina and the RPE. The RPE, but not the retina, can synthesize DHA from available ω3 precursors.

References

1 Fliesler SJ, Anderson RE: Chemistry and metabolism of lipids in the vertebrate retina. Prog Lipid Res 1983;22:79–131.

2 Benolken RM, Anderson RE, Wheeler TG: Membrane fatty acids associated with the electrical response in visual excitation. Science 1973;182:1253–1254.

3 Bourre J-M, Francois M, Youyou A, et al: The effects of dietary α-linolenic acid on the composition of nerve membrane, enzymatic activity, amplitude of electrophysiological parameters, resistance to poisons and performance of learning tasks in rats. J Nutr 1989; 119:1880–1890.
4 Neuringer M, Anderson GJ, Connor WE: The essentiality of n-3 fatty acids for the development and function of the retina and brain. Annu Rev Nutr 1988;8:517–541.
5 Uauy RD, Birch DG, Birch EE, et al: Effect of dietary omega-3 fatty acids on retinal function of very-low-weight neonates. Pediatr Res 1990;28:485–492.
6 Anderson RE, Maude MB: The effects of essential fatty acid deficiency on the phospholipids of photoreceptor membranes of rat retina. Arch Biochem Biophys 1972;151:270–276.
7 Galli C, Trzeciak HI, Paoletti R: Effects of dietary fatty acids on the fatty acid composition of the brain ethanolamine phosphoglyceride: Reciprocal replacement of n-6 and n-3 polyunsaturated fatty acids. Biochim Biophys Acta 1971;248:449–454.
8 Chen H, Wiegand RD, Koutz CA, et al: Docosahexaenoic acid increases in frog retinal pigment epithelium following rod photoreceptor shedding. Exp Eye Res 1992;55:93–100.
9 Bazan NG, Gordon WC, Rodriquez de Turco EB: Docosahexaenoic acid uptake and metabolism in photoreceptors: Retinal conservation by an efficient retinal pigment epithelial cells-mediated recycling process; in Bazan N, Marfi M, Tossano G (eds): Neurobiology of Essential Fatty Acids. New York, Plenum Press, 1992, pp 295–306.
10 Spector AA, Moore SA: Role of cerebromicrovascular endothelium and astrocytes in supplying docosahexaenoic acid to the brain; in Sinclair A, Gibson R (eds): Essential Fatty Acids and Eicosanoids. Champaign, American Oil Chemists' Society, 1992, pp 100–103.

Robert E. Anderson, PhD, MD, Cullen Eye Institute, Baylor College of Medicine,
One Baylor Plaza, Houston, TX 77030 (USA)

Galli C, Simopoulos AP, Tremoli E (eds): Fatty Acids and Lipids: Biological Aspects.
World Rev Nutr Diet. Basel, Karger, 1994, vol 75, pp 128–133

Local Synthesis and Targeting of Essential Fatty Acids at the Cellular Interface between Blood and Brain: A Role for Cerebral Endothelium and Astrocytes in the Accretion of CNS Docosahexaenoic Acid[1]

Steven A. Moore

Department of Pathology, The University of Iowa, Iowa City, Iowa, USA

Elongated, more highly polyunsaturated derivatives of linolenic acid (LNA), especially docosahexaenoic acid (DHA), accumulate in neurons of the brain [1]. This accretion of DHA in brain may depend upon local synthesis, a preformed external supply, or both. DHA accretion may, in addition, require mechanisms for targeting ω3 fatty acids to the brain and for limiting the egress of DHA from cerebral tissues. In the present work, these potential local mechanisms were studied in cell cultures of astrocytes, neurons, and cerebral microvascular endothelium grown individually or in co-culture combinations with one another. The findings suggest that endothelial cells of the blood-brain barrier cooperate with astrocytes in synthesizing DHA and in maintaining an ω3 fatty acid-enriched environment in the brain.

Methods

Cerebromicrovascular endothelial cells, astrocytes, and neurons were isolated from mouse or rat brains as described previously [2, 3]. The endothelial cell cultures were characterized and determined to be >95% pure by (a) light and electron microscopic

[1] This work was supported by National Institutes of Health Grants NS-01096, NS-27914, and NS-24621. Scientific collaborators in these studies include Arthur A. Spector, MD, Sean Murphy, PhD, and Gary Dutton, PhD. Linda Gorman, Elizabeth Yoder, and Erin Hurt performed the technical aspects of these studies.

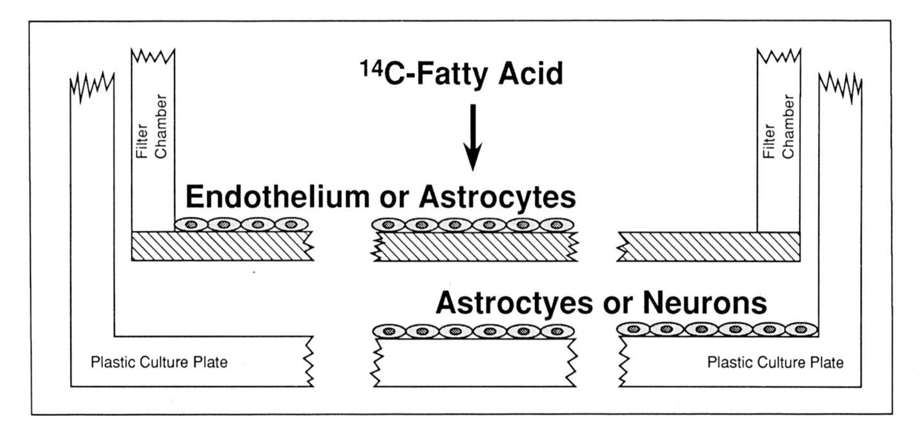

Fig. 1. In vitro model system for studying the transfer of ω3 fatty acids across the cerebral endothelium and examining the interactions among endothelium, astrocytes, and neurons.

appearance, (b) the presence of γ-glutamyltranspeptidase activity, (c) the uptake of DiI-Ac-LDL, and (d) Griffonia simplicifolia agglutinin histochemical staining followed by flow cytofluorometric analysis. Astrocyte cultures were routinely ≥95% type I astrocytes by immunohistochemical characterization. Neuronal cultures from cerebellum contained approximately 90% glutamatergic granule neurons, 5% other neurons, and 5% type I astrocytes.

For studies of individual cell types, astrocytes, neurons, or endothelium were grown in standard multiwell culture plates. For co-culture studies, astrocytes or endothelium were established on 25-mm filter chambers (Falcon) and cultures of cerebellar granule cell neurons or astrocytes were established in 6-well tissue culture plates as depicted in figure 1. The cells were incubated alone or in co-culture with either [1-^{14}C]-18:3ω3 or [1-^{14}C]-20:5ω3. Elongation and desaturation products in the medium and in cell lipids were determined by HPLC [2, 3].

Results and Discussion

Local Synthesis of Brain DHA

Although the essential polyunsaturated fatty acid DHA is highly enriched in brain neurons [1], it cannot be synthesized de novo by animal tissues and must ultimately be obtained from the diet. In addition to a direct dietary source, DHA can be synthesized from ω3 fatty acid presursors like LNA and eicosapentaenoic acid (EPA) through a process of fatty acid chain elongation and desaturation. The liver is a major site where shorter chain dietary ω3 fatty acid is converted to DHA, and one major source of DHA for the central nervous system is considered to be liver-derived DHA [4].

In addition to DHA formed in the liver, plasma also ordinarily contains ω3 fatty acid precursors [5], and circulating LNA and EPA may be additional

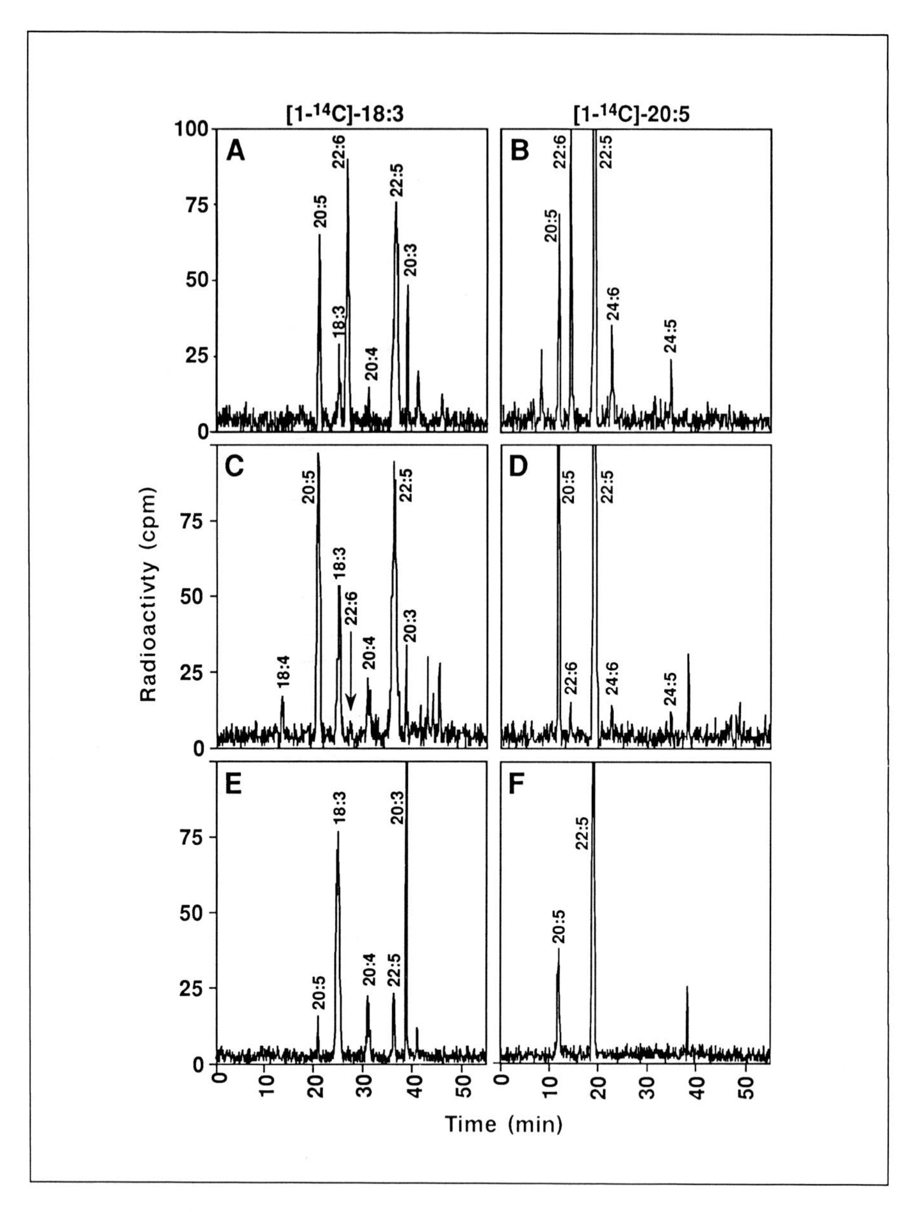

Fig. 2. Reverse phase HPLC chromatograms of cell lipid fatty acid methyl esters. *A* and *B* show the extensive elongation and desaturation of [1-^{14}C]-18:3ω3 and [1-^{14}C]-20:5ω3 and the production of substantial amounts of 22:6ω3 by astrocyte cultures after 24-hour incubations, respectively. *C* and *D* show the extensive elongation and desaturation of [1-^{14}C]-18:3ω3 and [1-^{14}C]-20:5ω3 after 24-hour incubations, but only minimal production of 22:6ω3 by endothelial cell cultures. *E* and *F* show that cerebellar neuronal cultures do not substantially desaturate either [1-^{14}C]-18:3ω3 and [1-^{14}C]-20:5ω3 after 72-hour incubations, but do elongate both fatty acids. These neuronal cultures do not produce appreciable amounts of 22:6ω3.

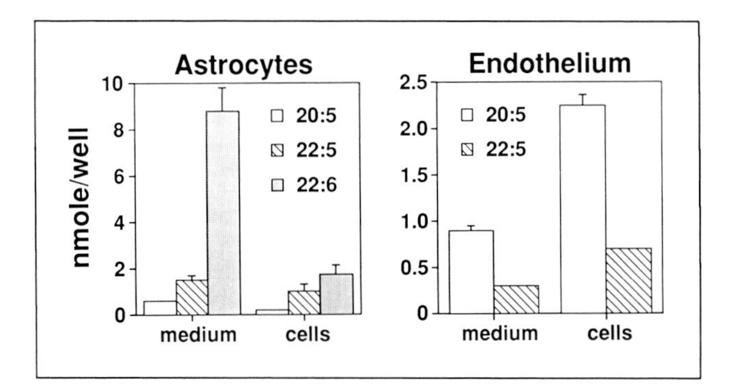

Fig. 3. Comparison of the elongation and desaturation products of [1-^{14}C]-18:3ω3 in the medium and cell lipids of astrocyte and endothelial cell cultures. Cell cultures were incubated with either [1-^{14}C]-18:3ω3 for 24 h. Bars represent means ± SE of triplicate cultures.

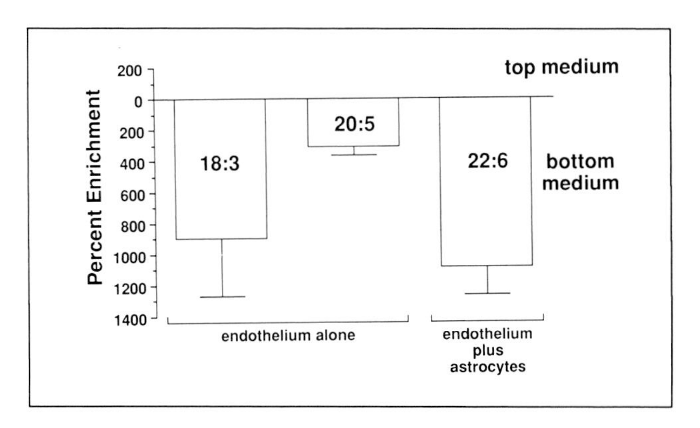

Fig. 4. Cell cultures were incubated with either [1-^{14}C]-18:3ω3 or [1-^{14}C]-20:5ω3 placed in the top medium of the filter chamber. After 96 h the bottom medium of endothelial cell cultures was highly enriched with the fatty acid originally placed in the top medium. In co-cultures of endothelium and astrocytes, the bottom medium was highly enriched with DHA produced by the astrocytes. Bars represent means ± SE of triplicate cultures.

sources of DHA for the brain. This pathway would require transfer of ω3 fatty acid precursors across the blood-brain barrier and conversion to DHA in the brain. It has been known for many years that brain tissue is capable of forming DHA from appropriate precursors [6], and the work presented here indicates that cerebral microvascular endothelium and astrocytes are capable of active elongation and desaturation of LNA and EPA (fig. 2). Cerebral endothelia produce and release primarily EPA, while astrocytes produce and release large amounts of DHA (fig. 2, 3). Neurons, on the other hand, are incapable of significant ω3 fatty acid elongation and desaturation, and are thus dependent upon a source of preformed DHA (fig. 2).

Targeting of ω3 Fatty Acids

In order for circulating LNA or EPA to be sources of DHA for the brain, they must first cross the blood-brain barrier. In an in vitro model of this barrier (fig. 1), LNA and EPA pass from the apical to the basal surface of the endothelium. Not only are these fatty acids transferred to the bottom medium, they are enriched in the medium facing the endothelial basolateral surface (fig. 4). This suggests that the cerebral endothelium may actively target ω3 fatty acids for release from its basal surface where these fatty acids can be enriched in the brain.

In co-cultures, the ω3 fatty acids targeted by endothelium to the bottom medium are readily taken up, elongated and desaturated by astrocytes to form DHA that in turn is released into the bottom medium. This astrocyte-derived DHA is enriched in the bottom medium to a degree similar to the LNA and EPA targeted into the bottom medium by endothelial cells cultured alone (fig. 4). In addition, these endothelial cell/astrocyte co-cultures produce greater total amounts of elongation and desaturation products than either cell type alone, as much as 50% more. These studies suggest that endothelial cells of the blood-brain barrier may cooperate with astrocytes by (a) providing astrocytes with ω3 fatty acid precursors, (b) increasing the degree of elongation and desaturation of ω3 fatty acids in the brain, and (c) blocking the egress of DHA from the brain.

DHA synthesized by astrocytes from radiolabeled precursors is readily incorporated by co-cultured neurons (data not shown). In fact, only when co-cultured with astrocytes do neurons contain radiolabeled DHA.

Conclusions

In composite the present studies and previously published work [2–4] support a dual pathway model for the accretion of CNS DHA that could utilize either DHA or its ω3 fatty acid precursors circulating in the blood. In one pathway the cerebral endothelium would take up preformed DHA and transfer it into the brain. In a second pathway cerebral endothelium would take up ω3 fatty acid precursors and target them preferentially into the brain, performing some elongation and desaturation in the process. Astrocytes would subsequently complete the conversion of precursors to DHA, releasing DHA for uptake by neurons. This second pathway would rely only on local processes in supplying DHA for the brain. In both pathways of this model, cerebral endothelium would perform the additional function of blocking the egress of

DHA out of the brain. In performing these local processes, cerebral endothelium and astrocytes may contribute positively to the high level of fatty acid desaturation necessary for normal neuronal function.

References

1 Salem N Jr, Kim HY, Yergey JA: Docosahexaenoic acid: membrane function and metabolism; in Simopoulos AP (ed): Health Effects of Polyunsaturated Fatty Acids in Seafoods. New York, Academic Press, 1986, pp 263–317.

2 Moore SA, Yoder E, Spector AA: Role of the blood-brain barrier in the formation of long-chain ω-3 and ω-6 fatty acids from essential fatty acid precursors. J Neurochem 1990;55:391–402.

3 Moore SA, Yoder E, Murphy S, et al: Astrocytes, not neurons, produce docosahexaenoic acid (22:6ω-3) and arachidonic acid (20:4ω-6). J Neurochem 1991;56:518–524.

4 Scott BL, Bazan NG: Membrane docosahexaenoate is supplied to the developing brain and retina by the liver. Proc Natl Acad Sci USA 1989;86:2903–2907.

5 Edelstein C: General properties of plasma lipoproteins and apolipoproteins; in Scanu AM, Spector AA (eds): Biochemistry and Biology of the Plasma Lipoproteins. New York, Dekker, 1986, pp 495–505.

6 Dhopeshwarkar GA, Subramanian C: Biosynthesis of polyunsaturated fatty acids in the developing brain. I. Metabolic transformations of intracranially administered 1-^{14}C linolenic acid. Lipids 1976;11:67–71.

Steven A. Moore, MD, PhD, Department of Pathology, The University of Iowa,
Iowa City, IA 52242 (USA)

Galli C, Simopoulos AP, Tremoli E (eds): Fatty Acids and Lipids: Biological Aspects.
World Rev Nutr Diet. Basel, Karger, 1994, vol 75, pp 134–138

Distribution, Processing and Selective Esterification of Essential Fatty Acid Metabolites in the Fetal Brain[1]

E. Yavin, P. Green

Department of Neurobiology, The Weizmann Institute of Science,
Rehovot, Israel

Critical Stages during Brain Development and Essential Fatty Acids

Under physiologic conditions, fetal brain development is regulated by the interaction of numerous genetic, environmental and nutritional factors [1]. The phase of neuronogenesis predates gliogenesis and myelinogenesis, the latter two processes constituting the phase of 'brain growth spurt' [2]. Neuronogenesis is both spatially and temporally regulated, and invariably in most mammals takes place during the intrauterine life. The rapid accumulation of long chain ω3 and ω6 polyunsaturated fatty acids (PUFAs) in the brain during prenatal and preweaning stages suggests that adequate supplies of PUFA may be critical for normal growth and function of the developing brain [3]. The importance of lipids in general and PUFAs in particular for normal cell function is well recognized [4]. The high content of docosahexaenoic acid (DHA) in brain tissue and retina are indicative of important roles this compound may fulfill [5, 6], in spite of the potential danger of being peroxidized and subject to free radical generation. Additionally, many physiologic and pathophysiologic processes such as vascular resistance, wound healing, reproduction, inflammation and allergy are modulated by PUFAs and their oxygenated derivatives [7].

[1] Supported by a grant from the Gulton Foundation, New York, N.Y.

The Fatty Acid Profile of the Developing Brain

Little is known about the developing brain during intrauterine life to selectively esterify PUFAs into phospholipids (PL) according to the polar head group composition. To categorize PUFA according to the polar head groups we have chosen to study the compositional details of the fatty acids (FA) in major PL classes at two distinct growth periods of the perinatal rat brain and compare it to the adult level. The period of day 17 to day 21 in utero is characterized by a rapid increase in both brain and body weight. At the end of this period, neurons that comprise the bulk of the cell population have ceased dividing and the entire brain enters a phase of active growth spurt primarily characterized by postnatal astroglial and oligodendroglial cell proliferation and differentiation. Though there is some variability between different brain regions, this process commences in the rat brain around the first week of postnatal life.

Cerebral hemispheres from 17- and 20-day gestational age rat fetuses, 8-day-old postnatal suckling pups and adult animals were collected and lipids extracted with hexane/isopropanol (3/2 by vol). The FA composition was determined by gas chromatography (GC) after TLC separation and trans-esterification of individual PL [8]. The time-dependent changes in the major PL-esterified FA and brain weight appear to increase at similar rates (fig. 1). Saturated FA constituted approximately 50% of total FAs at all the time points examined (data not shown). Principal changes were evident in the mono-unsaturated FA (MUFA) and PUFA species. MUFA increased more than 3-fold in the adult as compared to the 17-day fetus (7.11 and 1.97 μg/mg wet weight, respectively). In this respect, the FA composition of the fetal and early postnatal brain is quite similar. PUFA (ω3) increased 5-fold (from 0.59 to 3.31 μg/mg wet weight) and PUFA (ω6) increased 2-fold (from 1.39 to 3.28 μg/mg wet weight) in the adult compared to the fetus.

The distribution of the most ubiquitous FAs in selected PL species during early development was studied. As shown in figure 2, there is a general increase in the PUFA content within the PL species. Most notable is a substantial increase in ω3 PUFA content in PE, PE-plasmalogen (PE-pl) and PI lipids in the 8-day-old pups. These values are greater than those found in the adult brain. This may be explained by the lack of myelin in the 8-day-old rat brain. This is also supported by the gradual increase in the MUFA species in the adult, presumably because of myelin accretion.

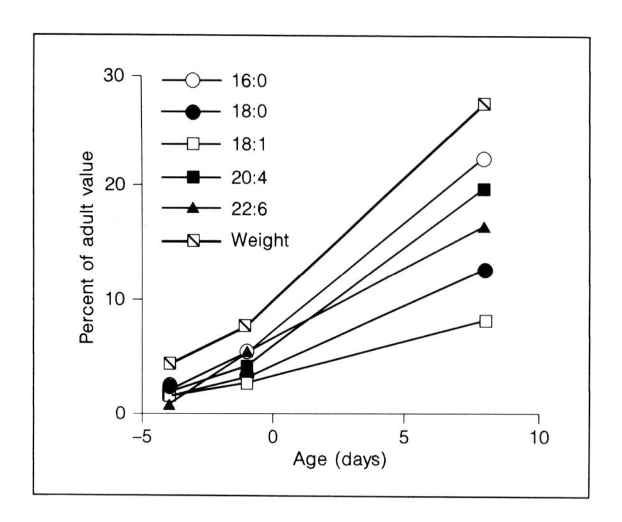

Fig. 1. Brain weight and total FA during development. Brains from 17- and 20-day fetuses, from 8th day postnatal pups and adult rats were removed, immediately frozen in liquid nitrogen and weighed. Brain lipids were extracted and transmethylated by BF_3 in methanol. Fatty acid methyl ester (FAME) analysis was performed on a 3400 Varian GC using a fused silica capillary column. Peak areas were integrated and plotted with the aid of a Varian Star Integrator computer package. Heneicosanoic acid (21:0) was used as internal standard for quantitation. FA acid peaks were identified by comparison to authentic standards. Parameters are expressed as percent of the corresponding adult value and are means of 2–4 brains.

Elongation and Desaturation of Linolenic and Linoleic Acids

To this date, most of the studies on the capability of brain tissue to elongate and desaturate linolenic acid (LNA) and linoleic acid (LA) were carried out in postnatal animals [9–12]. There is still some controversy as to whether the fetal brain is capable to elongate-desaturate the two EFA and selectively incorporate the resulting long-chain PUFA into individual lipids.

To further examine this unsettled question, and test whether long-chain PUFAs are selectively esterified according to the polar head group composition, tracer amounts of either [1^{14}C]LNA or [1-^{14}C]LA were intracranially injected into 19- to 20-day-old rat fetuses and their metabolism examined for up to 20 h [8, 13]. A rapid disappearance of the free precursors, with apparent half-lifes of 60 and 40 min for LNA and LA respectively, was noticed. One hour after LNA injection, 32.3 and 14.3% of the total brain radioactivity was found in the neutral glycerides (NG) and PL fractions, respectively. After 20 h, PL radioactivity attained a level of 75%. An increase in the amount of labeled DHA which resided predominantly in the PE and PE-pl was observed. A similar pattern of incorporation into NG and PL fraction was observed after

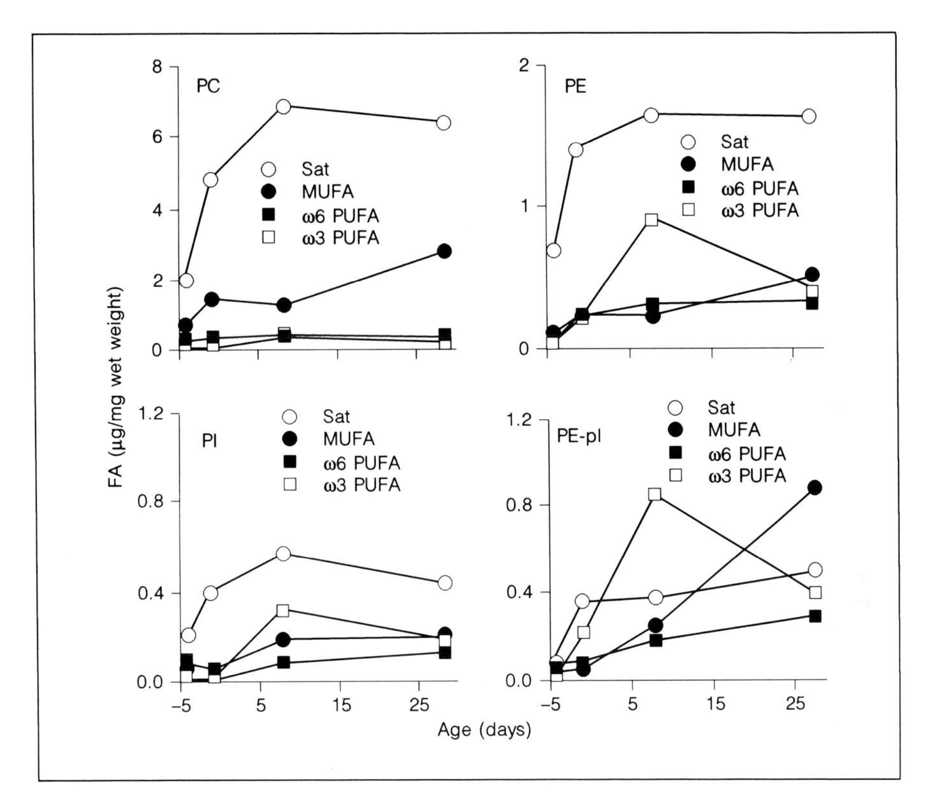

Fig. 2. Phospholipid FA accretion during development. The major PLs, phosphatidylcholine (PC), phosphatidylethanolamine (PE), phosphatidylinositol (PI) and phosphatidylserine (PS), as well as the ethanolamine plasmalogen (PE-pl), were separated in two directions using for the first direction a mixture of chloroform:methanol:methylamine (40%) (130:70:30 by vol) and for the second direction chloroform:acetone:methanol:acetic acid:water (100:40:20:30:10 by vol) with exposure to HCl fumes between the two runs. Spots were identified after spraying with dichlorofluorescein and scraped off the plates. FA were transmethylated and the FAMEs separated and identified as described in the legend to figure 1. FA are expressed as μg/mg wet weight and are means of 4 brains.

the administration of [1-^{14}C]LA, except that the resulting labeled arachidonic acid (AA) was esterified preferentially in PI.

Selective esterification of PUFAs according to polar head group composition was also notable after direct intracerebral injection of [^{3}H]DHA or [^{3}H]AA. However, in contrast to LNA and LA uptake, radioactively labeled DHA and AA were esterified into PL almost without delay. Of the total [^{3}H]DHA esterified, 29.2 and 20% were esterified in the ethanolamine and choline PL respectively. There was 12 and 40% label from [^{3}H]AA esterified in the ethanolamine and choline PL respectively. Inositol PL labeling by [^{3}H]AA was 6- to 8-fold higher than that seen in the presence of DHA. Within the neutral lipids, diacyl- and triacylglycerol species were labeled 14.3 and 12.6% by [^{3}H]DHA and 8.9 and 9.3% by [^{3}H]AA, suggesting a potential metabolic function for these lipids in the developing brain. The exact mechanism for the preferential labeling pattern is not clear. It could possibly relate to the selectivity of the acyl transferase and transacylase systems to esterify the 2-position of the glycerol moiety according to the number of double bonds and polar head group characteristics.

The present results indicate that the near-term fetal rat brain has the capacity to take up, convert, and selectively esterify essential FAs and their long chain PUFAs into PLs. While the liver is probably the major source for brain long-chain PUFA in the postnatal animal, the presence of an independent metabolic pathway in the brain during intrauterine life may be an additional mechanism for modeling brain PUFA profile.

References

1 Dobbing G: Vulnerable periods of brain development; in Elliott K, Knight J (eds): Lipids, Malnutrition and the Developing Brain. Amsterdam, Associated Scientific Publishers, 1972, pp 9–20.
2 Dobbing J, Sands J: Comparative aspects of the brain growth spurt. Early Hum Dev 1979;3:79–83.
3 Salem N, Ward GR: Are ω-3 fatty acids essential nutrients for mammals? World Rev Nutr Diet. Basel, Karger, 1993, vol 72, pp 128–147.
4 Dratz EA, Deese AJ: The role of docosahexaenoic acid (22:6ω3) in biological membranes: Examples from photoreceptors and model membrane bilayers; in Simopoulos AP, Kifer RR, Martin RE (eds): Health Effects of Polyunsaturated FAs in Seafoods. New York, Academic Press, 1986, pp 319–351.
5 Neuringer M, Connor WE, Lin DS, Barstad L, Luck SJ: Biochemical and functional effects of prenatal and postnatal ω-3 fatty deficiency in retina and brain in rhesus monkey. Proc Natl Acad Sci USA 1986;83:4021–4025.
6 Salem N Jr, Kim H-Y, Yergey JA: Docosahexaenoic acid: Membrane function and metabolism; in Simopoulos AP, Kifer RR, Martin RE (eds): Health Effects of Polyunsaturated FAs in Seafoods. New York, Academic Press 1986, pp 263–317.
7 Wolfe LS: Eicosanoids: Prostaglandins, thromboxanes, leukotrienes, and other derivatives of carbon-20 unsaturated fatty acids. J Neurochem 1982;38:1–14.
8 Green P, Yavin E: Elongation-desaturation of linolenic and linoleic acids in rat foetal brain in vivo; in Drevon CA, Baksaas I, Krokan HE (eds): Proceedings of the Symposium Omega-3 Fatty Acids, Metabolism and Biological Effects. Basel, Birkhäuser, 1993, pp 131–138.
9 Scott BL, Bazan NG: Membrane docosahexaenoate is supplied to the developing brain and retina by the liver. Proc Natl Acad Sci USA 1989;86:2903–2907.
10 Nouvelot A, Delbart C, Bourre JM: Hepatic metabolism of dietary alpha-linolenic acid in suckling rats, and its possible importance in polyunsaturated fatty acid uptake by the brain. Ann Nutr Metab 1986;30:316–323.
11 Hassam AG, Crawford MA: The differential incorporation of labelled linoleic, γ-linolenic, dihomo-γ-linolenic and arachidonic acids into the developing rat brain. J Neurochem 1976; 27:967–968.
12 Cohen SR, Bernsohn J: The in vivo incorporation of linolenic acid into neuronal and glial cells and myelin. J Neurochem 1978;30:661–669.
13 Green P, Yavin E: Elongation, desaturation and esterification of essential fatty acids by fetal rat brain in vivo. J Lipid Res, 1993;34:2099–2107.

E. Yavin, PhD, Department of Neurobiology, The Weizmann Institute of Science,
Rehovot 76100 (Israel)

Galli C, Simopoulos AP, Tremoli E (eds): Fatty Acids and Lipids: Biological Aspects.
World Rev Nutr Diet. Basel, Karger, 1994, vol 75, pp 139–141

Effect of Diet on Term Infant Cerebral Cortex Fatty Acid Composition

E.C. Jamieson, K.A. Abbasi, F. Cockburn, J. Farquharson, R.W. Logan, W.A. Patrick

University Department of Child Health, Royal Hospital for Sick Children, Glasgow, UK

Fossil records show that mammals first appeared on earth about 250 million years ago and placental mammals only in the last 100 million years. About 40 million years ago a tailless primitive ape, possibly related to man's ancestors, first appeared. In the last 10 thousand years man has learned to cultivate plants and domesticate animals for food consumption. It is only in the last 130 years that man or woman in many 'developed' countries has opted out of being truly mammalian and has modified milk from other mammals or created synthetic plant-based 'milks' to feed newborn human infants. This uncontrolled experiment could have major consequences for the development of the infant brain.

During the first year of life the infant brain weight increases from 350 to 1,100 g and of the 750 g weight gain about 85% is cerebrum of which 50–60% of the solid matter is lipid. We do not yet understand the processes which control fatty acid accretion but we have previously demonstrated that cortical total phospholipid fatty acid composition in term and preterm infants is considerably influenced by their dietary fat intake [1]. Cerebral cortical neuronal membrane glycerophospholipids are comprised predominantly (95%) by phosphatidylcholine (PC), phosphatidylethanolamine (PE) and phosphatidylserine (PS). The effect of feeding different milk formulae on the fatty acids attached to the phosphatides of PS and PE in human infant brain are now described.

Methods

Fatty acids were measured in total lipid extracts and in PS and PE fractions isolated by two-dimensional thin-layer chromatography of lipid extracts obtained from the cerebral cortex from 17 breast- and formula-fed, first-year, crib-death infants. Formula-fed infants were 6 infants fed SMA and 5 fed Cow and Gate or Ostermilk (CGOST), the last 2 having similar fatty acid compositions at the time of the study. SMA contained 1.5% of the essential fatty acid α-linolenic acid (C18:3ω3) whereas there was only 0.4% in the CGOST milks. Only the 6 human milk-fed infants received any docosahexaenoic acid (DHA, C22:6ω3).

Results

There were no significant differences in saturated and monounsaturated fatty acids among the three groups of infants. Examination of the total phospholipid results showed that breast-fed infants had significantly greater concentrations of DHA (C22:6ω3) in their cerebral cortex than both of the formula-fed groups (breast 9.7% vs. SMA 7.5% and CGOST 6.6%). PE in the cerebral cortex of breast-fed infants contain significantly more C22:6ω3 than both of the formula-fed groups (17.7% vs. 13.4% SMA and 11.6% CGOST). The deficiency of the ω3 series was substituted by the ω6 series so that the total ω6 fatty acid content in the formula-fed infants' brains exceeded that in the breast-fed (34.5% vs. 39.6% SMA and 42.9% CGOST). PS in the cerebral cortex of the breast-fed infants contained significantly more C22:6ω3 than both formula-fed groups (23.5% vs. 19.3% SMA and 14.4% CGOST), again the deficit in ω3 series was made up by substitution of the ω6 series (22.8% vs. 27.1% SMA and 31.0% CGOST). It would appear that docosapentaenoic acid (DPA, C22:5ω6) largely substitutes for the deficiency of DHA in neuronal membrane phospholipid.

Discussion

After birth, neuronal membranes and retinal photoreceptor cells derive most of their phospholipid DHA from diet and liver synthesis rather than from subcutaneous fat reserves and neuronal synthesis. Neither the liver nor the retinal and neuronal cells can synthesize DHA if there is no fat tissue reserve or inadequate dietary supply of precursor essential fatty acid (α-linolenic), or if the synthetic enzymes required are not yet activated or are inactivated by excess ω6 fatty acids [2]. When the diet after birth is human milk there is available a supply of ready-synthesized DHA and arachidonic acid (AA). In many infant formulae there is little or no DHA or AA and in some there is insufficient essential fatty acid precursor (α-linolenic and linoleic acids) to allow for their synthesis. Evidence from animal studies suggests that retinal function and learning ability are permanently impaired if there is a failure in the accumulation of sufficient DHA during development [3].

There is need for a good prospective long-term (60+ years) study on the effect of feeding artificial formulae to the newborn on subsequent health and brain function. There will always be a need for a small number of infants to have artificial formulae when their own mother's milk or milk from a wet nurse or milk bank is unavailable. In the meantime, given our knowledge of the effects of formula feeding on the chemical composition of the brain there should be every effort made to keep the number of formula-fed infants to an absolute minimum. The short-term effect on efficiency of synaptic transmission and longer term effects on neuronal membrane integrity brought about by these neuronal membrane phospholipid fatty acid changes need urgent study. It may be that the introduction of synthetic formulae for the feeding of the newborn human infant is a major step forward in the evolutionary process but it is more likely to be a major retrograde step.

References

1 Farquharson J, Cockburn F, Patrick WA, Jamieson EC, Logan RW: Infant cerebral cortex phospholipid fatty-acid composition and diet. Lancet 1992;340:810–813.

2 Farquharson J, Cockburn F, Patrick WA, Jamieson EC, Logan RW: Effect of diet on infant subcutaneous tissue triglyceride fatty acid and implications for cerebral cortex composition. Arch Dis Child 1993;69:589–593.

3 BNF Task Force Report: Unsaturated Fatty Acids – Nutritional and Physiological Significance. London, Chapman & Hall, 1992, pp 63–67.

E.C. Jamieson, BSc, University of Glasgow, Royal Hospital for Sick Children,
Glasgow G3 8SJ (UK)

Galli C, Simopoulos AP, Tremoli E (eds): Fatty Acids and Lipids: Biological Aspects.
World Rev Nutr Diet. Basel, Karger, 1994, vol 75, pp 142–143

Summary Statement: PUFA and Natural Antioxidants

R. Muggli

This session was co-chaired by *S.N. Meydani and R. Muggli,* and presentations were made by Drs. *Meydani, Muggli, M.S. Nenseter, G. Hornstra, L.B.M. Tijburg, J.P. Allard,* and *A. Petroni.*

Fats are vulnerable to attack by oxygen free radicals and must be protected by antioxidants. Vitamin E, which is a fat-soluble compound, has been recognized as being especially important in preventing damage to fats by attack from oxygen free radicals. While there is clear evidence for a requirement of vitamin E in connection with the amount and degree of unsaturation of PUFA in the diet, the exact amount of vitamin E needed to compensate for this increased demand caused by PUFA in the diet has not been systematically investigated in man. Thus adequate blood levels of vitamin E are essential in clinical trials designed to prove the beneficial effects of fish and fish oil as cardioprotectors and anti-inflammatory agents. It would be unfortunate if a beneficial effect of ω3 LCPUFA should be missed because of insufficient antioxidant protection. Such a case was presented, attributing the decreased immune response in man after fish or fish oil, despite a reduction in PGE_2, in part due to reduced tissue and plasma vitamin E.

0.6 mg vitamin E (*d*-α-tocopherol) per gram of PUFA (linoleic acid) is generally regarded as the critical value of protection. The amount required to protect higher unsaturated PUFA increases approximately linearly with the number of double allylic positions. A theoretically derived formula for its calculation was proposed.

During the symposium, it was discussed that the difficulty in establishing antioxidant requirements roots in the multiplicity of natural antioxidants, and in the dependency of the results on the particular biological parameter investigated. In addition, differing in vitro and in vivo findings may make it difficult to assess the relevance of laboratory findings.

Besides vitamin E, other known (vitamin C, β-carotene, uric acid, etc.) and unknown natural antioxidants most certainly play a role in defending against oxygen damage. In olive oil, 3,4-dihydroxyphenyl ethanol was identified as one of several phenolic antioxidants with antithrombotic affects in vitro.

While high intakes of PUFA in animals and man can lead to reduced plasma and tissue vitamin E levels, and to symptoms of relative vitamin E deficiency, such as creatinuria, erythrocyte peroxidizability and elevated lipid peroxide markers, lipid peroxidation, as measured by breath ethane and pentane output, was not increased in volunteers after supplementation of 5.3 g EPA and DHA in the form of ethyl esters for 6 weeks.

Particularly disturbing is the nagging question as to whether PUFA, fish oil in particular, will increase the susceptibility of LDL to oxidative changes. In contrast to one study which found the oxidation resistance of LDL significantly reduced following fish oil consumption, another study found that 5 g marine ω3 fatty acids per day for 4 months had no effect on the susceptibility of LDL to oxidation. In view of the pivotal role of oxidatively modified LDL as a possible pathogenetic factor in atherosclerosis, this discrepancy should be urgently resolved.

Galli C, Simopoulos AP, Tremoli E (eds): Fatty Acids and Lipids: Biological Aspects.
World Rev Nutr Diet. Basel, Karger, 1994, vol 75, pp 144–148

Modification of Low Density Lipoprotein in Relation to Intake of Fatty Acids and Antioxidants

Marit S. Nenseter[a], *Vivi Volden*[b], *Serena Tonstad*[c], *Ola Gudmundsen*[b], *Leiv Ose*[c], *Christian A. Drevon*[a]

Institutes for [a]Nutrition Research and [b]Biology, University of Oslo, and [c]The Lipid Clinic, National Hospital, Oslo, Norway

Several lines of evidence suggest that low density lipoprotein (LDL) must be oxidatively modified to be rapidly taken up by macrophages in the arterial wall [1] (fig. 1). Uptake of modified LDL via the scavenger receptor may lead to foam cell formation, fatty streaks and atherosclerotic plaques. The purpose of the present study was to examine the effects of oral supplementation with (a) marine ω3 polyunsaturated fatty acids, (b) β-carotene and (c) cholesterol on the susceptibility of LDL to oxidation.

Subjects, Study Design and Methods

ω3 Polyunsaturated Fatty Acids. In a pilot study, normolipidemic subjects (n = 2) were supplemented with 5.1 g marine ω3 fatty acids (K85; Pronova Biocare AS, Norway) per day for 4 months, whereas control subjects (n = 2) received a similar amount of corn oil. The capsules contained 4 mg α-tocopherol/g fatty acids. Normolipidemic, young women (n = 8) were supplemented with 5.4 g marine ω3 fatty acids (EPAX6000; Martens AS, Norway) per day for 6 weeks. LDL was isolated at baseline, and after 3 and 6 weeks. The control group (n = 8) received an equal amount of oil with a fatty acid pattern similar to an ordinary Norwegian diet. The ω3 fatty acid and control capsules contained 2 and 0.5 mg tocopherols/g fatty acids, respectively.

β-Carotene. Hypercholesterolemic, postmenopausal women were supplemented with 30 mg β-carotene (Roche, Switzerland) per day (n = 16 subjects) or placebo capsules (n = 15 subjects) for 10 weeks. LDL was isolated from all subjects before and after treatment.

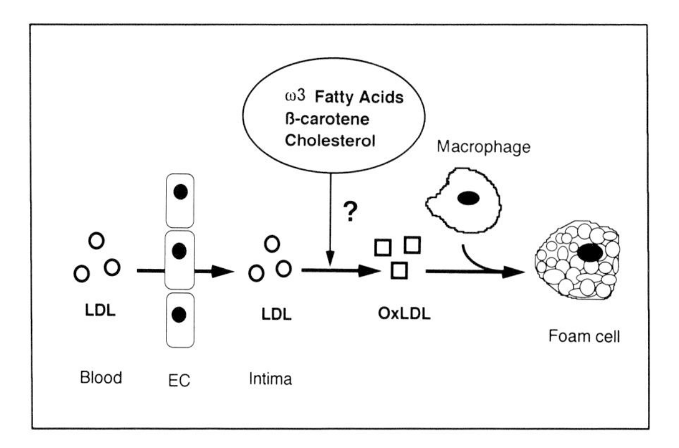

Fig. 1. Schematic model of the oxidatively modified LDL (OxLDL) hypothesis. For more details, see the introduction. EC = Endothelial cells.

Cholesterol. Rabbits were fed a diet containing 2% (wt/wt) cholesterol for 3 weeks (n = 4), whereas control rabbits received an unsupplemented diet (n = 3). LDL was isolated after 3 weeks of supplementation.

Oxidative Modification of LDL. LDL was subjected to Cu^{2+}-catalyzed lipid peroxidation [2], and the extent of oxidation was assessed by measuring the formation of conjugated dienes [2], lipid peroxides and thiobarbituric acid-reactive substances [2], changes in the fatty acid pattern [2], relative electrophoretic mobility in agarose gels, and uptake of lipoproteins by J774 macrophages [3].

Results and Discussion

ω3 Polyunsaturated Fatty Acids

The amounts of ω3 fatty acids in plasma were significantly increased, and ω6 fatty acids significantly reduced in the fish oil groups, whereas no changes were measured in the fatty acid patterns in the control groups. All the methods used to assess the extent of oxidation suggested that ω3 fatty acid supplementation had no measureable effect on the susceptibility of LDL to modification, as compared to baseline values or to the control groups. The results are in agreement with the study by Alessandrini et al. [4], in a group of patients with type IIB hyperlipidemia, showing no significant differences betwen LDL from patients supplemented with ω3 fatty acids and olive oil. It has been reported that α-tocopherol protects LDL against lipid peroxidation [5]. Our results showed that the levels of α-tocopherol and vitamin C in plasma did not change significantly during the treatment period, and there were no significant differences between the control and the fish oil groups. This may partly explain our results. Furthermore, Fisher et al. [6] have shown that feeding fish oil to

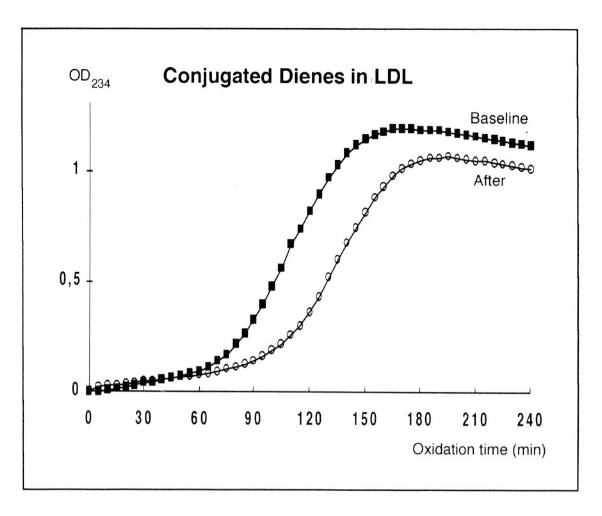

Fig. 2. Continuous monitoring of optical density at 234 nm (OD_{234}) in LDL from one subject in the β-carotene group, at baseline and after treatment. 25 μg LDL was subjected to Cu^{2+}-catalyzed oxidation (1.67 μM) for 4 h at 37 °C.

humans causes a reduction in oxygen free radical formation by their neutrophils and monocytes. This may contribute to counteract the expected enhanced susceptibility of the ω3 fatty acid-enriched LDL particles to lipid peroxidation.

β-Carotene

In vitro studies have shown conflicting results on the effect of β-carotene to protect LDL from modification [7]. In addition, it has been reported that oral supplementation with β-carotene had no protective effect in a group of smokers [8]. In the present study, hypercholesterolemic, postmenopausal women were supplemented with β-carotene. Figure 2 shows the continuous monitoring of conjugated dienes in LDL from one subject in the β-carotene group, at baseline and after treatment. From this analysis we calculated the lag time for the onset of the oxidation, the oxidation rate, and the maximum amount of conjugated dienes formed. These data, as well as the amount of lipid peroxides formed and the relative electrophoretic mobility, indicated that supplementation of β-carotene had no protective effect on the susceptibility of LDL to modification. That does not rule out that β-carotene may have a protective effect in vivo.

Cholesterol

It is well documented that hypercholesterolemia is a risk factor for development of coronary heart disease. Furthermore, Lavy et al. [9] have demonstrated enhanced in vitro oxidation of LDL from hypercholesterolemic patients. We have shown that cholesterol feeding of rabbits resulted in markedly increased uptake of LDL in the liver endothelial and Kupffer cells, giving them a foam cell-like morphology [10]. We speculated that LDL was modified in the hypercholesterolemic rabbits, to a form recognized by the scavenger/oxidized LDL receptor in these cell types. Copper-catalyzed oxidation of LDL suggested that LDL from the hypercholesterolemic rabbits were more susceptible to lipid peroxidation than LDL from the normolipidemic rabbits. Furthermore, LDL from the normolipidemic rabbits contained a 4-fold higher amount of α-tocopherol than LDL from the hypercholesterolemic rabbits. The findings support our hypothesis and suggest that the liver endothelial and Kupffer cells play a protective role by removing atherogenic lipoproteins from the circulation.

Conclusions

Marine polyunsaturated ω3 fatty acid supplementation had no measurable effect on the susceptibility of LDL to lipid peroxidation, suggesting that ω3 fatty acids do not render the LDL particle more atherogenic. β-Carotene supplementation had no protective effect on the susceptibility of LDL to lipid peroxidation in our systems. Cholesterol feeding and hypercholesterolemia in rabbits increased the susceptibility of LDL to lipid peroxidation. Taken together, our data suggest that dietary modification of LDL may have implications for the development of atherosclerosis.

References

1 Witztum JL, Steinberg D: Role of oxidized low density lipoprotein in atherogenesis. J Clin Invest 1991;88:1785–1792.

2 Parthasarathy S, Khoo JC, Miller E, et al: Low density lipoprotein rich in oleic acid is protected against oxidative modification: Implications for dietary prevention of atheosclerosis. Proc Natl Acad Sci USA 1990;87:3894–3898.

3 Nenseter MS, Rustan AC, Lund-Katz S, et al: Effect of dietary supplementation with n-3 polyunsaturated fatty acids on physical properties and metabolism of low density lipoprotein in humans. Arterioscler Thromb 1992;12:369–379.

4 Bittolo-Bon G, Cazzolato G, Alessandrini P, et al: Effects of concentrated DHA and EPA supplementation on LDL peroxidation and vitamin E status in type IIB hyperlipidemic patients; in Drevon CA, Baksaas I, Krokan HE (eds): Omega-3 Fatty Acids: Metabolism and Biological Effects. Basel, Birkhäuser, 1993, pp 51–58.

5 Dieber-Rotheneder M, Puhl H, Waeg G, et al: Effect of oral supplementation with *D*-α-tocopherol on the vitamin E content of human low density lipoproteins and resistance to oxidation. J Lipid Res 1991;32:1325–1332.

6 Fisher M, Upchurch KS, Johnson MH, et al: Effects of dietary fish oil supplementation on polymorphonuclear leukocyte inflammatory potential. Inflammation 1986;10:387–391.

7 Jialal I, Norkus EP, Cristol L, et al: β-Carotene inhibits the oxidative modification of low density lipoprotein. Biochim Biophys Acta 1991;1086:134–138.

8 Princen HMG, van Poppel G, Vogelzang C, et al: Supplementation with vitamin E but not β-carotene in vivo protects low density lipoprotein from lipid peroxidation in vitro: Effect of cigarette smoking. Arterioscler Thromb 1992;12:554–562.
9 Lavy A, Brook GJ, Dankner G, et al: Enhanced in vitro oxidation of plasma lipoproteins derived from hypercholesterolemic patients. Metabolism 1991;40:794–799.
10 Nenseter MS, Gudmundsen O, Roos N, et al: The role of liver endothelial and Kupffer cells in clearing low density lipoprotein from blood in hypercholesterolemic rabbits. J Lipid Res 1992;33: 867–877.

Marit S. Nenseter, PhD, Institute for Nutrition Research, University of Oslo,
PO Box 1046, Blindern, N-0316 Oslo (Norway)

Galli C, Simopoulos AP, Tremoli E (eds): Fatty Acids and Lipids: Biological Aspects.
World Rev Nutr Diet. Basel, Karger, 1994, vol 75, pp 149–154

Peroxidation of Low Density Lipoproteins and Endothelial Phospholipids: Effect of Vitamin E and Fatty Acid Composition

G. Hornstra[a], *G.S. Oostenbrug*[a], *R.C.R.M. Vossen*[b]

Departments of [a]Human Biology and [b]Biochemistry, Limburg University, Maastricht, The Netherlands

Fatty acid (per)oxidation of low density lipoproteins (LDL) is likely to play an important role in atherogenesis. Oxidatively modified LDL can be taken up in an uncontrolled way by macrophages via so-called 'scavenger receptors'. As a result, these macrophages turn into cholesterol-laden foam cells, thereby forming fatty streaks which, ultimately, can develop into atherosclerotic plaques [1]. Fatty acids are more sensitive to peroxidation the higher their degree of unsaturation. Since the polyunsaturated fatty acid (PUFA) composition of LDL is a reflection of that of the diet [2], the oxidation resistance of LDL can be expected to decrease upon the consumption of a diet rich in PUFA. Fish oil is rich in highly unsaturated fatty acids and, therefore, the effect of dietary fish oil and of vitamin E consumption on the oxidizability of LDL is of interest.

Lipid peroxidation may also promote cholesterol influx into the arterial wall via a direct damaging effect on endothelial cells, which are rich in both ω6 as well as ω3 PUFA. In addition, endothelial injury may give rise to thrombus formation and, therefore, the fatty acid composition of endothelial membrane phospholipids may be of importance in atherogenesis and thrombogenesis via an effect on the sensitivity of endothelial cells to peroxidation.

Effect of Fish Oil and Vitamin E Consumption on LDL Peroxidation in vitro

Seven healthy male volunteers were given a dietary supplement containing 2 g ω3 PUFA from fish oil/day for a period of 3 weeks. Four of them also consumed an extra 300 mg of vitamin E/day. Four additional men, who did not receive any supplement, served as controls. Before and after the supplementation period, blood was sampled from a forearm vein and LDL was isolated by ultracentrifugation. After extensive dialysis to remove the anticoagulant (EDTA), 0.25 g/l LDL was oxidized in vitro with $CuCl_2$ (final concentration 1.66 μmol/l) at 37 °C. Fatty acid peroxidation was monitored by continuously measuring conjugated diene formation in a spectrophotometer at 234 nm and converting extinction units into amounts of dienes, using a molar extinction coefficient of 29,500 $M^{-1} \cdot cm^{-1}$. After addition of the copper ions, the antioxidants present in the LDL particles inhibit oxidation of the LDL PUFA. After these protective antioxidants have been fully consumed, conjugated dienes are formed, which is reflected by a rise in absorption at 234 nm. The longer the time between $CuCl_2$ addition and the start of conjugated diene formation (the lag time), the higher the ratio between the amounts of antioxidants and oxidizable PUFA present in the LDL particles. When corrected for changes occurring in the control group, fish oil consumption was associated with a reduction of the lag time, which was not only prevented by the additional vitamin E intake, but even reversed to a prolongation. The difference in response between both supplemented groups was statistically significant (fig. 1). Once all antioxidants had been consumed, there were no differences between the experimental groups anymore, neither in the increase in total amounts of conjugated dienes formed (which was significant for both groups (fig. 1b) and resulted from the higher amounts of PUFA present in the LDL particles), nor in the rate of diene formation, the changes of which were not statistically significant (fig. 1c). This study strongly indicates that an adequate vitamin E intake protects LDL particles from enhanced peroxidation after an increased consumption of ω3 PUFA. However, because of the relatively small number of volunteers involved in this study, and considering the inconsistent results with respect to the effect of fish oil consumption on the oxidation resistance of LDL [2; see also Nenseter and Alessandrini, this volume], more studies are needed before a final conclusion can be drawn.

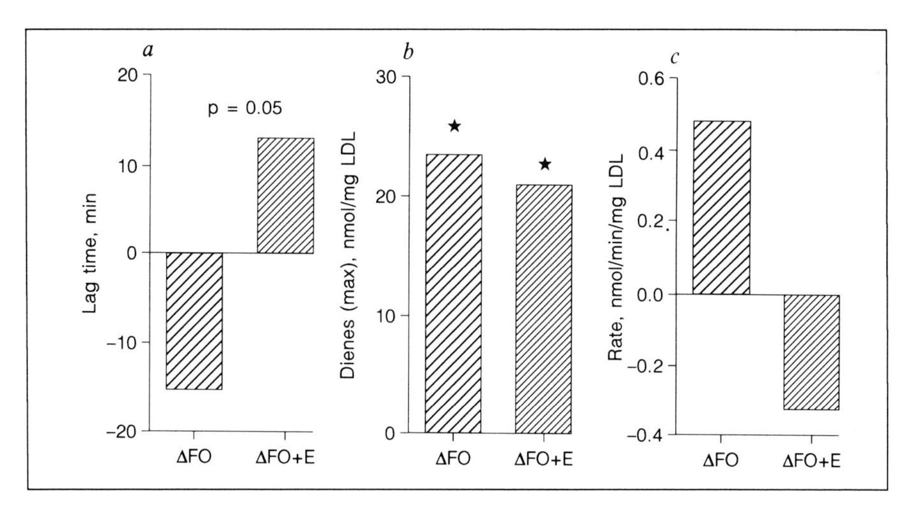

Fig. 1.a–c Effect of the daily consumption of 2 g ω3 PUFA as fish oil, either (FO + E) or not (FO) in combination with 300 mg vitamin E, for 3 weeks on the oxidation resistance of LDL. Results are given as the difference (Δ) with respect to an unsupplemented control group. * Difference with control: $p<0.05$. For further explanations see text.

Fatty Acid Composition and the Sensitivity of Endothelial Phospholipids to Peroxidation

Endothelial cells obtained from human umbilical veins were cultured for several passages in a standard medium containing 20% human serum. Cells were detached from the culture plates, washed extensively and total lipids were extracted. Subsequently, phospholipids were isolated by thin-layer chromatography and used for the preparation of liposomes, which were subjected to peroxidation by the combined action of copper sulfate and hydrogen peroxide.

In a first study, the fatty acid compositions of the liposomes were determined by quantitative gas-liquid chromatography (GLC) before and after peroxidation for 80 min, and peroxidative fatty acid loss was calculated. Only fatty acids having two or more double bonds appeared to be peroxidized. In addition, a significant, positive, rectilinear relationship was observed between the sensitivity of these fatty acids to peroxidation and their number of double bonds, docosahexaenoic acid (DHA, 22:6ω3) being the most sensitive.

In a next series of studies, the fatty acid composition of endothelial cell phospholipids was modulated by growing the cells for three passages (7–8 days) in a culture medium to which various fatty acids were added as their sodium salts (200 μM final concentration). As a result of this procedure, major alterations were induced in the fatty acid compositions of the endothelial

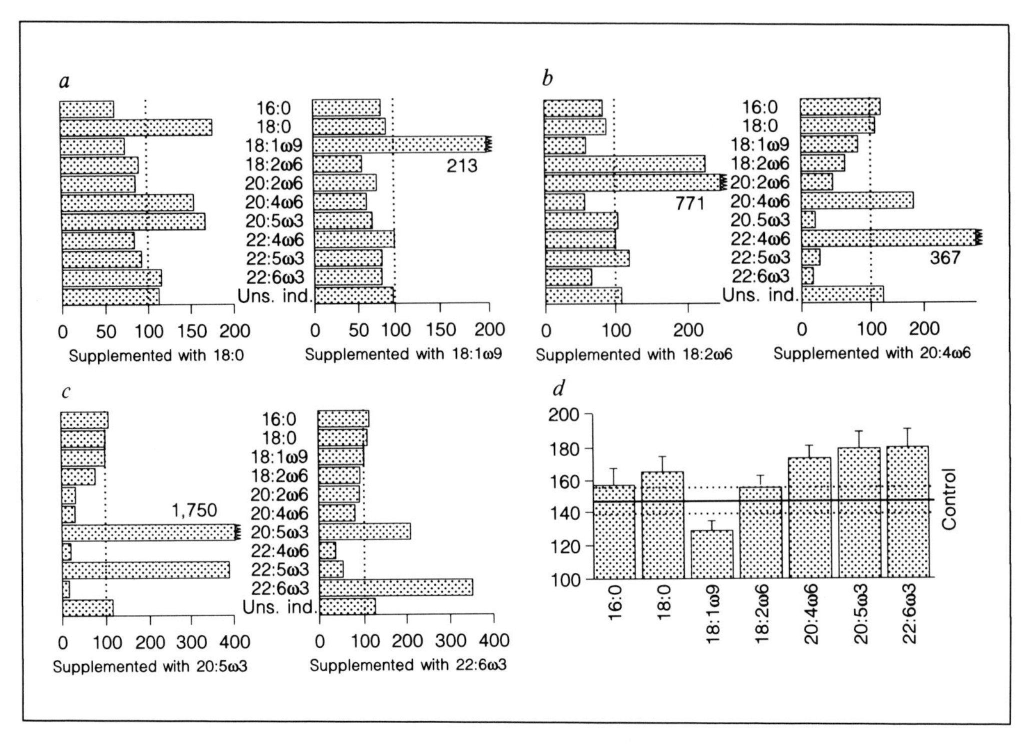

Fig. 2. Fatty acid composition (*a–c*, % of control cells, indicated by broken lines) and unsaturation indices (*d*, mean + SEM) of phospholipids, isolated from endothelial cells cultured in media supplemented with various fatty acids (200 μ*M*). For further explanations see text.

phospholipids (fig. 2), which in some cases caused a significant difference in the unsaturation index (the sum of the percentage of each fatty acid in a phospholipid preparation, multiplied by the number of double bonds contained in that fatty acid). The cells showed normal morphology, growth rate and von Willebrand factor staining when cultured under these conditions [3].

Endothelial phospholipid peroxidation was monitored in a double beam spectrophotometer, using difference spectroscopy, which greatly improved the sensitivity of the conjugated diene measurement. The method was further optimized by the use of tandem cuvettes, which rules out disturbances of the different UV spectra caused by changes in the concentrations of liposomes, $CuSO_4$ or H_2O_2. This method, which enables a convenient, sensitive, and continuous registration of conjugated diene formation in an aqueous environment, gave results which compared quite well with those obtained with discontinuous and more laborious methods such as the measurement by GLC of peroxidative fatty acid loss, and the quantitation of lipid hydroperoxide formation, using an iodometric assay [4].

The higher the total amounts of PUFA present in the liposomes prepared from the fatty acid-modified endothelial cells, the higher the extent of lipid peroxidation. Growing the cells in an oleate-enriched medium caused the lowest degree of peroxidation, whereas conjugated diene formation was highest in liposomes prepared from endothelial cells grown in a linoleate-enriched medium. When the cells were cultured in a medium enriched with eicosapentaenoic acid (EPA, 20:5ω3) or DHA (22:6ω3), the contents of these fatty acids and the unsaturation indices of endothelial phospholipids were significantly increased. Conjugated diene formation in these liposomes, however, was hardly different from that of unmodified control cells. This suggests that under the prevailing experimental conditions these highly unsaturated ω3 fatty acids do not induce an increased sensitivity to phospholipid peroxidation and may be protected from extensive peroxidation.

Fatty Acid Modification of Endothelial Cells: Direct Cell-Damaging Effect of Peroxidative Stress

Cells were grown in fatty acid-supplemented media and then subjected to short-term peroxidative stress. In one study, DNA strand breaking (a marker of early oxidative injury) was measured by ethidium bromide fluorescence [5]. Incubation of the confluent monolayers with 0–75 μM H_2O_2 for 30 min, progressively increased the number of DNA strand breaks. However, no differences were observed between the cells grown in the media enriched with the various fatty acids. When the H_2O_2 dose was increased to the millimolar range, cell lysis occurred, which was measured by the release of preabsorbed ^{51}Cr. Again no differences were observed between the cells grown in the various fatty acid-supplemented media.

PUFA Peroxidation, Deleterious or Beneficial?

These studies suggest that, in contrast to short-term cell culture experiments [6, 7], long-term culturing of endothelial cells in fatty acid-supplemented media enables the cells to develop effective homeostatic control, which is likely to neutralize the functional consequences of the occurring fatty acid changes. Fatty acid peroxidation results in the formation of peroxides which are able to cause cellular damage. However, formation of hydroperoxides from PUFA is tightly coupled to their reaction with free radicals, which are more aggressive than lipid peroxides. The results of our 'nutrition experiments in vitro' suggest that the beneficial radical scavenging potential of PUFA may be at least as important as the potential adverse effects of peroxide formation. This has been suggested before by Dormandy [8], and recent work by Sosenko et al. [9–11] is in line with this concept. They demonstrated that feeding ω6 or ω3 PUFA to pregnant rats significantly increases the PUFA content of the lungs of their pups but, nonetheless, protects these rats from oxygen toxicity [9–11]. However, much more research needs to be done to improve our understanding of the functional balance between radical scavenging and peroxide formation.

References

1 Parthasarathy S, Steinberg D, Witzum JL: The role of oxidized low density lipoproteins in the pathogenesis of atherosclerosis. Annu Rev Med 1992;43:219–225.
2 Nenseter MS, Rustan AC, Lund-Katz S, et al: Effect of dietary supplementation with n-3 polyunsaturated fatty acids on physical properties and metabolism of low density lipoproteins in humans. Arterioscler Thromb 1992;12:369–379.
3 Vossen RCRM, van Dam-Mieras MCE, Lemmens PJMR, et al: Membrane fatty acid composition and endothelial cell functional properties. Biochim Biophys Acta 1992;1083:243–251.
4 Vossen RCRM, van Dam-Mieras MCE, Hornstra G, et al: Continuous monitoring of lipid peroxidation by measuring conjugated diene formation in an aqueous liposome suspension. Submitted.
5 Lorenzi M, Montisano DF, Toledo S, et al: High glucose induces DNA damage in cultured human endothelial cells. J Clin Invest 1986;77:322–325.
6 Hart CM, Tolson JK, Block E: Supplemental fatty acids alter lipid peroxidation and oxidant injury in endothelial cells. Am J Physiol 1991;260:L481–L488.
7 Spitz DR, Kinter MT, Kehrer JP, et al: The effect of monounsaturated and polyunsaturated fatty acids on oxygen toxicity in cultured cells. Pediatr Res 1992;32:366–372.
8 Dormandy TL: Biological rancidification. Lancet 1969;ii:684–688.
9 Sosenko IRS, Innis SM, Frank L: Polyunsaturated fatty acids and protection of newborn rats from oxygen toxicity. J Pediatr 1988;112:630–637.
10 Sosenko IRS, Innis SM, Frank L: Menhaden fish oil, n-3 polyunsaturated fatty acids and protection of newborn rats from oxygen toxicity. Pediatr Res 1989;25:399–404.
11 Sosenko IRS, Innis SM, Frank L: Intralipid increases lung polyunsaturated fatty acids and protects newborn rats from oxygen toxicity. Pediatr Res 1991;30:413–417.

G. Hornstra, Department of Human Biology, Limburg University, PO Box 616, NL-6200 MD Maastricht (The Netherlands)

Galli C, Simopoulos AP, Tremoli E (eds): Fatty Acids and Lipids: Biological Aspects.
World Rev Nutr Diet. Basel, Karger, 1994, vol 75, pp 155–161

Interaction of ω3 Polyunsaturated Fatty Acids and Vitamin E on the Immune Response

Simin Nikbin Meydani

Nutritional Immunology Laboratory, USDA Human Nutrition Research Center on Aging at Tufts University, Boston, Mass., USA

Epidemiological studies indicate that consumption of fish and fish oil is associated with a low rate of coronary heart disease mortality [1]. In age-associated inflammatory diseases such as rheumatoid arthritis, consumption of fish oil improves clinical parameters such as swollen joints, tenderness and pain, and grip strength [2, 3]. Furthermore, fish oil supplementation has been shown to be beneficial in the treatment of inflammatory bowel disease [4] and lupus erythematosus.

The immune system and its protein and lipid mediators have been implicated in pathogenesis of atherosclerotic and inflammatory diseases. The prevalence of these diseases increases with age while the function of the immune system is altered as a function of age.

The beneficial effects of fish and fish oils are attributed to their ω3 polyunsaturated (PUFA) content, in particular to eicosapentaenoic acid (EPA) and docosahexaenoic acid (DHA). EPA and DHA are incorporated in the cell membrane where they can influence membrane fluidity, receptor function, enzyme activity and production of eicosanoids. Eicosanoids play an important role in control of the inflammatory and immune responses. The incorporation of highly unsaturated fatty acids such as EPA and DHA into cellular membrane potentiates their peroxidizability and increases the requirement for antioxidant nutrients. Vitamin E is the major fat-soluble, chain-breaking antioxidant in biological membranes and protects membrane PUFA from lipid peroxidation [5]. Earlier studies have shown that the requirement for vitamin E increases with increased consumption of PUFA [6]. Adequate level of vitamin E is needed for normal function of the immune cells. We have, therefore, studied the effect of fish oil and its interaction with vitamin E on immune response in context of aging.

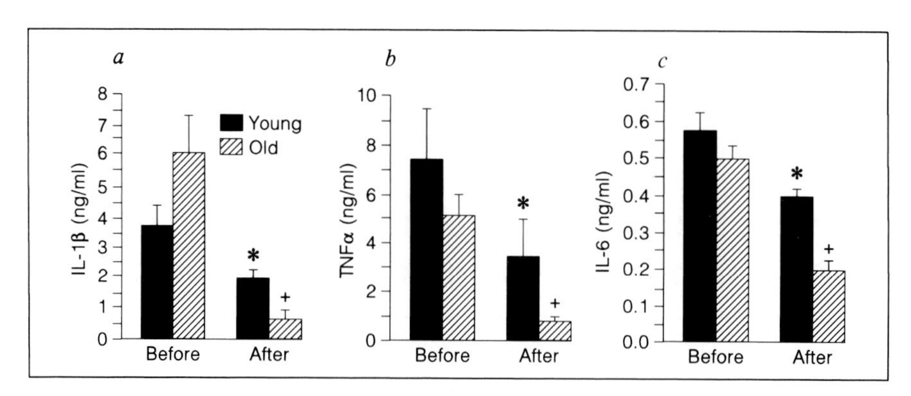

Fig. 1. Effect of ω3 PUFA supplementation (2.4 g/day for 3 months) on IL-1β (*a*), TNF_{α} (*b*) and IL-6 (*c*) production in young and older women (mean ± SEM, n = 6). *Significantly lower than before values at $p < 0.05$. +Significantly lower than young at $p < 0.05$.

Fish Oil and Immune Response

The effect of dietary ω3 PUFA on cytokine production and lymphocyte proliferation was studied in 6 young (20- to 33-year-old) healthy women who supplemented their typical American diet with ω3 PUFA contained in six capsules of Pro-Mega™ daily for 12 weeks [7]. Each subject received 1.68 g of EPA, 0.72 g of DHA, 0.6 g of other fatty acids and 6 IU of vitamin E per day [7]. Heparinized blood was collected on 2 consecutive days from the young women during the follicular phase of their menstrual cycle and 2 consecutive days from older women. The means of the two measurements were used for statistical comparison. Mitogenic lymphocyte proliferation, induction of cytokines, and PGE_2 formation were determined. A significant increase in plasma EPA and a significant decrease in arachidonic acid (AA)/EPA ratio was observed; the changes were more dramatic in the older women than in the young women. Total number of white blood cells and the percentage of mononuclear cells did not change with age or as a result of treatment.

Proinflammatory cytokine interleukin (IL)-1β, tumor necrosis factor $(TNF)_{\alpha}$ and IL-6 production was not significantly different between the young and older women prior to ω3 PUFA supplementation (fig. 1). ω3 PUFA supplementation significantly decreased production of these cytokines in both groups. The decrease was more pronounced in the older women with a significantly lower production of these proinflammatory cytokines noted after 3 months of ω3 PUFA supplementation (fig. 1). In this study fish oil consumption significantly decreased two T-cell-mediated functions, i.e. IL-2 production and

Table 1. Effect of ω3 PUFA supplementation on IL-2 production and mitogenic response of PBMC from young and older women (mean ± SEM, n = 6)

Parameter	Young		Older	
	before	after	before	after
IL-2, U/ml	88 ± 28	38 ± 20	60 ± 30*	22 ± 9⁺
Mitogenic response to PHA, cpm × 10^3	84 ± 16	78 ± 9	54 ± 6	34 ± 8⁺

Con A-stimulated IL-2 production and PHA-stimulted mitogenic response was measured as described [20]. Data represent corrected counts per minute, which is the cpm of stimulated culture minus cpm of unstimulated culture. There was no difference in cpm of unstimulated cultures between young and older women or before and after supplementation with ω3 fatty acids.

* Significantly lower than young at $p < 0.05$.

⁺ Significantly lower than before values at $p < 0.05$.

mitogenic response to phytohemagglutin (PHA) in older women (table 1). These responses were lower in the older women prior to fish oil consumption compared to those of the young subjects and were further significantly reduced after fish oil supplementation (table 1).

We have also found suppression of immune responsiveness when subjects consumed diets enriched with fish-derived ω3 PUFA [8]. Twenty-two subjects (> 40 years old) were fed a diet approximating that of the current American diet (14.1 en% saturated fatty acids (SFA), 14.5 en% monounsaturated fatty acids (MUFA), 6.1 en% ω6 PUFA, 0.8 en% ω3 PUFA, 147 mg cholesterol/1,000 cal) for 6 weeks after which time they consumed (11 subjects in each group) one of the two low-fat, low-cholesterol, high-PUFA diets based on National Cholesterol Education Panel (NCEP) Step-2 recommendations (4.0–4.5 en% SFA, 10.8–11.6 en% MUFA, 10.3–10.5 en% PUFA, 45–61 mg cholesterol/1,000 cal) for 6 months. One of the NCEP Step-2 diets was enriched in fish-derived ω3 PUFA (0.54 en% or 1.23 g/day EPA and DHA) (referred to as low-fat, high-fish diet) and the other low in fish-derived ω3 PUFA (0.13 en% or 0.27 g/day EPA and DHA) (referred to as low-fat, low-fish diet). Measurements of in vivo and in vitro indices of immune responses were taken following each dietary period (three measurements at each time point for in vitro tests). We used Multi-test CMI™ to measure delayed-type hypersensitivity skin test (DTH), an in vivo indicator of cell-mediated immunity.

This study shows that long-term feeding of a low-fat, high-fish diet significantly decreases the percentage of helper T cells whereas the percentage of suppressor cells is increased [8]. We also found that mitogenic responses to Con

Table 2. Effect of low-fat, high-fish and low-fat, low-fish diets on immune response of humans (mean ± SE, n = 10)

Immune indices	Baseline	Low fat, high-fish	Baseline	Low-fat, low-fish
Mitogenic response to Con A, cpm × 10^3	24.30 ± 3.10	19.40 ± 3.20*	34.70 ± 3.40	44.40 ± 6.10*
DTH (induration index, mm)	15.70 ± 2.40	8.60 ± 1.90*	14.70 ± 2.80	15.40 ± 3.20
IL-1β, ng/ml	5.70 ± 1.02	3.44 ± 0.78*	4.28 ± 1.49	6.93 ± 1.64*
TNF, ng/ml	13.90 ± 2.59	8.98 ± 2.04*	8.18 ± 1.92	9.56 ± 1.70
IL-6, ng/ml	1.14 ± 0.13	0.75 ± 0.08*	0.75 ± 0.20	0.75 ± 0.26

For mitogenic response, 10^6 PBMC/ml were stimulated with 50 μg/ml Con A for 72 h. For IL-1β and TNF, 5 × 10^6 FBMC/ml were cultured with 1 ng/ml LPS and 20 organisms/cell of *Staphylococcus epidermis* respectively for 24 h. For IL-6, 10^6 PBMC/ml were cultured with 10 μg/ml Con A for 48 h. DTH was measured using Multi-test CMI. For details of methods, see Meydani et al. [8].

*Significantly different from baseline at $p < 0.05$ by Student's paired t test.

A and DTH as well as the production of cytokines IL-1β, TNFα and IL-6 by mononuclear cells were significantly reduced following the low-fat, high-fish diet (table 2). In contrast, long-term feeding of a low-fat, low-fish diet enriched with plant-derived PUFA increased peripheral blood mononuclear cells (PBMC) mitogenic response to Con A, IL-1β and TNFα production but had no effect on DTH, or TNFα and IL-6 production (table 2).

The mechanism of ω3 PUFA-induced immunosuppression is not understood. Three possible mechanisms have been suggested: (1) changes in eicosanoid production; (2) increased free radical formation and lipid peroxidation resulting in increased tocopherol requirement, and (3) changes in early activation events such as protein kinase C activation, phosphotidylinositol turnover and calcium mobilization. The second possibility is discussed below.

Role of Vitamin E in Fish Oil-Induced Immunological Changes

Vitamin E plays a regulatory role in the maintenance of the immune response. Severe and marginal vitamin E deficiencies in several species of animals, as well as in humans [9, 10] is associated with a decrease in T-cell-mediated functions. Furthermore, higher than RDA levels of vitamin E have been shown to enhance the immune response in aged mice [11] and humans [12].

As mentioned above, fish oil feeding increases the requirement for tocopherol. Therefore, it is possible that the immunosuppressive effect of fish oil is due to compromised vitamin E status and that supplementation with adequate levels of vitamin E will reverse fish-oil-induced immunosuppression.

Vitamin E is the major fat-soluble, chain-breaking antioxidant in biological membranes and protects membrane PUFA from lipid peroxidation [13]. Early studies, using biochemical measures, have demonstrated that the requirement for this antioxidant increases with increased consumption of PUFA [14]. In fact, in the early studies, vitamin E deficiency was produced by feeding cod liver oil [15]. The ability of vitamin E to protect membranes from oxidative damage is dependent on the magnitude and duration of oxidative stress. We have found that in young and old mice supplemented with different levels of vitamin E, fish-oil-fed mice had lower plasma and tissue α-tocopherol levels compared to mice fed corn oil or coconut oil [16].

Consumption of 10 g of MaxEPA fish oil for 4 weeks was shown to increase the basal level of lipid peroxides in plasma of smokers and resulted in a higher peroxidative modification of LDL and uptake by macrophages [17]. Supplementation of the subjects with 400 mg vitamin E decreased some of the cigarette-induced oxidation of LDL and uptake by macrophages [17]. Meydani et al. [18] have reported increases in plasma lipid peroxides in healthy young and older women after taking 6 capsules of fish oil (providing 2.4 g/day EPA and DHA) with their meals daily for a 3-month period. The older women showed a significantly greater increase in plasma EPA and DHA, a greater increase in PUFA/SFA ratio, and had higher plasma lipid peroxides compared to young women. In addition, even though vitamin E/ml of plasma did not change, fish oil supplementation significantly decreased the vitamin E/(EPA+DHA) ratio in plasma. As mentioned above in our studies using diets based on NCEP recommendations, the low-fat, high-fish diet decreased the indices of immune response measured while the low-fat, low-fish diet increased or had no effect on these indices. It is interesting to note that in this study the low-fat, high-fish diet decreased the plasma E/PUFA ratio ($p < 0.04$) while the low-fat, low-fish diet had no significant effect on E/PUFA ratio, indicating that tocopherol status might be important in determining the immunological effect of fish oil.

This is supported by our preliminary observation in mice where in vitro addition of vitamin E to spleen cells from mice fed fish oil increased mitogenic response by 300% as opposed to a 100% increase observed when vitamin E was added to cells from corn-oil-fed mice [19]. This is further supported by the study of Kremer et al. [20] where supplementation of humans with 200 mg tocopherol reversed the depressed mitogenic response induced by feeding 15 g/day of a fish oil supplement.

Conclusion

The ω3 PUFA have profound effects on the ability of immune cells to produce protein and lipid mediators and to exhibit their effector function. The immunological changes induced by ω3 PUFA contributes to the anti-inflammatory and antiatherosclerotic effects of fish oil. This beneficial effect, however, is dampened by its potential for increasing free radical formation and lipid peroxidation in the absence of adequate antioxidant protection. Increasing the proportion of fish in the typical American diet or consuming fish oil supplements without adequate antioxidant protection may result in in vivo peroxidation of ω3 PUFA and thereby reduce its beneficial cardiovascular, antiaggregatory and anti-inflammatory effects. Therefore, the potentially advantageous effects of dietary ω3 PUFA may be offset by the deleterious consequence of free radical formation and lipid peroxidation including a decrease in T-cell-mediated immunity (proliferative response to T-cell mitogens and DTH). These risks associated with the intake of ω3 PUFA may be minimized without compromising its beneficial effects by the intake of appropriate levels of an antioxidant such as vitamin E.

Further studies are needed to determine the adequate level of vitamin E when ω3 PUFA intake is increased so that the beneficial effects of fish oil can be achieved without its adverse effect on the antioxidant defense system and T-cell-mediated function.

References

1 Dyerberg J, Bang HO: Haemostatic function and platelet polyunsaturated fatty acids in Eskimos. Lancet 1979;ii:433–435.
2 Kremer JM, Jubiz W, Michalek A, et al: Fish oil fatty acid supplementaion in active rheumatoid arthritis. A double-blinded, controlled, crossover study. Ann Intern Med 1987;106: 497–503.
3 Lee TH, Hoover RL, Williams JD, et al: Effect of dietary enrichment with eicosapentaenoic and docosahexaenoic acid on in vitro neutrophil and monocyte leukotriene generation and neutrophil function. N Engl J Med 1985;321:1217–1224.
4 Stenson WF, Cort D, Rodgers J, et al: Dietary supplementation with fish oil in ulcerative colitis. Ann Intern Med 1992;116:609–614.
5 Virella G, Kilpatrick JM, Rugeles MT, et al: Depression of humoral responses and phagocytic functions in vivo and in vitro by fish oil and eicosapentaenoic acid. Clin Immunol Immunopathol 1989;52:257–270.
6 Santoli D, Zurier RB: Prostaglandin E precursor fatty acids inhibit human IL-2 production by a prostaglandin E-dependent mechanism. Immunology 1989;143:1303–1309.
7 Meydani SN, Endres S, Woods MN, et al: Oral (n-3) fatty acid supplementation suppresses cytokine production and lymphocyte proliferation: Comparison between young and older women. J Nutr 1991;121:547–555.
8 Meydani SN, Lichenstein AH, Cornwall S, et al: Immunological effects of National Cholesterol Education Panel (NCEP) Step-2 diets with and without fish-derived (n-3) fatty acid enrichment. J Clin Invest 1993;92:105–113.

9 Bendich A, Machlin LJ: Safety of oral intake of vitamin E. Am J Clin Nutr 1988;48:612–619.
10 Kowdley KV, Meydani SN, Cornwall SC, et al: Reversal of depressed T-lymphocyte function with repletion of vitamin E deficiency. Gastroenterology 1992;102:2139–2142.
11 Meydani SN, Meydani M, Verdon CP, et al: Vitamin E supplementation suppresses prostaglandin E_2 synthesis and enhances the immune response of aged mice. Mech Ageing Dev 1986;34:191–210.
12 Meydani SN, Barklund PM, Liu S, et al: Vitamin E supplementation enhances cell-mediated immunity in healthy elderly subjects. Am J Clin Nutr 1990;52:557–563.
13 Chow CK: Vitamin E and oxidative stress. Free Radic Biol Med 1991;11:215–222.
14 Witting LA: Vitamin E-polyunsaturated lipid relationship in diet and tissues. Am J CLin Nutr 1974;27:952–959.
15 MacKenzie GG, MacKenzie JB, McCollum EV: Uncomplicated vitamin E deficiency in rabbit and its relation to the toxicity of cod liver oil. J Nutr Sci 1941;21:225–234.
16 Meydani SN, Shapiro AC, Meydani M, et al: Effect of age and dietary fat (fish, corn and coconut oils) on tocopherol status of C57BL/6Nia mice. Lipids 1987;22:345–350.
17 Harats D, Dabach Y, Hollander G, et al: Fish oil ingestion in smokers and nonsmokers enhances peroxidation of plasma lipoproteins. Atherosclerosis 1991;90:127–139.
18 Meydani M, Natiello F, Goldin B, et al: Effect of long-term fish oil supplementation on vitamin E status and lipid peroxidation in women. J Nutr 1991;121:484–491.
19 Shapiro AC, Wu D, Meydani SN: Eicosanoids derived from arachidonic and eicosapentaenoic acids inhibit T-cell proliferative response. Prostaglandins 1993;45:229–240.
20 Kremer JR, Schoene N, Dougless LW, et al: Increased vitamin E intake restores fish-oil-induced suppressed blastogenesis of mitogenic-stimulated T lymphocytes. Am J Clin Nutr 1991;54:896–902.

Simin Nikbin Meydani, DVM, PhD, Nutritional Immunology Laboratory,
USDA Human Nutrition Research Center on Aging at Tufts University,
711 Washington Street, Boston, MA 02111 (USA)

Galli C, Simopoulos AP, Tremoli E (eds): Fatty Acids and Lipids: Biological Aspects.
World Rev Nutr Diet. Basel, Karger, 1994, vol 75, pp 162–165

Effect of ω3 Fatty Acids and Vitamin E Supplements on Lipid Peroxidation Measured by Breath Ethane and Pentane Output: A Randomized Controlled Trial[1]

Johane P. Allard, Dawna Royall, Regina Kurian, Reto Muggli, Khursheed N. Jeejeebhoy

Department of Medicine, University of Toronto, Ont., Canada;
F. Hoffmann-La Roche Ltd., Basel, Switzerland

The purpose of this randomized, double-blind, controlled trial, was to investigate in humans (1) the effect of ω3 polyunsaturated fatty acid supplementation on lipid peroxidation measured by breath alkane output (BAO) and by plasma malondialdehydes (MDA) and (2) whether vitamin E supplementation can suppress lipid peroxidation.

Method

Eighty volunteers were randomized (using random number tables) into four groups in a double-blind fashion to receive for 6 weeks either menhaden oil ethyl esters (8 caps/day eicosapentaenoic acid (EPA) = 3.062 g; docosahexaenoic acid (DHA) = 2.262 g; total ω3 = 6.26 g) or its placebo (p), olive oil ethyl esters (8 caps/day = 6.08 g monoenes) with *d*-α-tocopheryl acetate (vitamin E) (3 caps/day = 900 IU) or its placebo. Breath collections [1] and biochemical measurements were performed in the morning, in the fasted state, at baseline (week 0) and at weeks 3 and 6 of supplementation. Blood was withdrawn for analysis of plasma carotenes, α- and γ-tocopherol [2], vitamin C [3], phospholipid composition [4] and MDA [5].

All group data are expressed as mean ± SEM. ANOVA for repeated measures was used to make comparisons between groups at baseline, week 3 and week 6. When ANOVA was significant, differences between groups were examined by the Newman-Keuls technique.

[1] Supported by a grant from F. Hoffmann-La Roche Ltd.

Table 1. Population

	ω3, vitamin E	ω3, vitamin Ep	ω3p, vitamin E	ω3p, vitamin Ep
Number	18	17	19	18
Age, years	32 ± 2	33 ± 3	32 ± 2	31 ± 2
BMI, kg/m^2	24.7 ± 0.5	25.0 ± 1.1	23.3 ± 0.4	25.4 ± 0.7
% pill taken				
ω3/p	91.1 ± 1.4	84.7 ± 2.5*	91.6 ± 1.4	91.6 ± 2.3
Vitamin E/p	98.4 ± 0.6	95.2 ± 2.0	98.2 ± 0.7	96.1 ± 1.6

* $p < 0.05$.

Table 2. Vitamin E levels

	α-Tocopherol, μmol/l		
	week 0	week 3	week 6
ω3, vitamin E	21.1 ± 0.9	42.1 ± 2.3*	42.9 ± 2.8*
ω3, vitamin Ep	21.5 ± 1.4	21.7 ± 1.5	20.8 ± 1.8
ω3p, vitamin E	24.1 ± 0.8	53.4 ± 5.3*	54.6 ± 4.0*
ω3p, vitamin Ep	22.8 ± 2.1	23.3 ± 1.6	23.5 ± 1.7

* $p < 0.001$.

Results

Of the 80 volunteers, 72 completed the study. The four groups were comparable at the baseline (table 1) for age, body mass index (BMI) as well as for BAO, plasma MDA, α-tocopherol, β-carotene, vitamin C and phospholipid composition (EPA, DHA, AA). Although compliance was excellent, it was significantly lower in one group.

During supplementation, there was a significant increase in plasma α-tocopherol levels in those receiving vitamin E but not in the placebo groups (table 2). Other antioxidant vitamins remained stable throughout the study. There was also a significant increase in EPA and DHA as well as a decrease in AA in plasma phospholipids when supplemented with ω3 fatty acids (table 3).

Table 3. Plasma phospholipid composition

	Week 0	Week 3	Week 6
Eicosapentaenoic acid, % comp.			
ω3, vitamin E	0.99 ± 0.13	8.38 ± 0.69*	8.85 ± 0.56*
ω3, vitamin Ep	1.07 ± 0.12	8.39 ± 0.77*	8.31 ± 0.89*
ω3p, vitamin E	1.13 ± 0.09	1.07 ± 0.13	0.99 ± 0.09
ω3p, vitamin Ep	1.25 ± 0.15	1.16 ± 0.15	1.15 ± 0.11
Docosahexaenoic acid, % comp.			
ω3, vitamin E	3.65 ± 0.24	7.31 ± 0.36*	8.11 ± 0.26*
ω3, vitamin Ep	4.15 ± 0.42	7.86 ± 0.44*	8.32 ± 0.52*
ω3p, vitamin E	3.95 ± 0.24	3.49 ± 0.20	3.32 ± 0.10
ω3p, vitamin Ep	4.22 ± 0.38	4.39 ± 0.36	3.70 ± 0.24
Arachidonic acid, % comp.			
ω3, vitamin E	14.23 ± 0.71	11.63 ± 0.47**	11.02 ± 0.58*
ω3, vitamin Ep	13.22 ± 0.70	12.07 ± 0.56**	11.13 ± 0.46
ω3p, vitamin E	14.45 ± 0.55	13.90 ± 0.57	13.67 ± 0.49
ω3p, vitamin Ep	12.66 ± 0.48	13.32 ± 0.61	12.25 ± 0.40

* $p < 0.01$; ** $p < 0.05$.

Table 4. Lipid peroxidation indices

	Week 0	Week 3	Week 6
*Ethane, pmol/kg/min**			
ω3, vitamin E	12.2 ± 1.4	9.2 ± 1.1	10.3 ± 1.4
ω3, vitamin Ep	11.0 ± 1.0	12.0 ± 1.4	11.7 ± 1.5
ω3p, vitamin E	11.3 ± 1.0	11.5 ± 1.4	12.5 ± 1.7
ω3p, vitamin Ep	11.2 ± 1.1	13.8 ± 1.5	12.8 ± 1.6
*Pentane, pmol/kg/min**			
ω3, vitamin E	6.3 ± 1.2	4.9 ± 0.6	5.4 ± 0.8
ω3, vitamin Ep	5.3 ± 0.7	4.6 ± 0.6	5.8 ± 1.0
ω3p, vitamin E	5.1 ± 0.6	5.0 ± 0.5	4.6 ± 0.5
ω3p, vitamin Ep	7.5 ± 1.6	8.0 ± 1.7	5.5 ± 1.1
MDA, nmol/ml			
ω3, vitamin E	2.12 ± 0.28	3.96 ± 0.56**	3.75 ± 0.37**
ω3, vitamin Ep	2.27 ± 0.38	4.17 ± 0.87**	3.85 ± 0.67**
ω3p, vitamin E	2.50 ± 0.30	2.34 ± 0.29	1.78 ± 0.21
ω3p, vitamin Ep	2.57 ± 0.37	2.06 ± 0.24	1.95 ± 0.29

* No significant differences between groups; ** $p < 0.01$.

During the study, there was no significant change in breath ethane or pentane output among the four groups (table 4). However, plasma MDA levels increased significantly during ω3 fatty acid supplementation compared to the other groups at week 3 and week 6. There was no significant effect of vitamin E supplementation on these lipid peroxidation indices.

Conclusion

The results of the present study demonstrated that ω3 fatty acid supplementation increases lipid peroxidation measured by plasma MDA. However, breath ethane and pentane output did not change significantly. In addition, vitamin E did not suppress lipid peroxidation during ω3 fatty acid supplementation.

References

1 Lemoyne M, Van Gossum A, Kurian R, et al: Breath pentane analysis as an index of lipid peroxidation: A functional test of vitamin E status. Am J Clin Nutr 1987;46:267–272.
2 Hess D, Keller HE, Oberlin B, et al: Simultaneous determination of retinol, tocopherols, carotenes and lycopene in plasma by means of high-performance liquid chromatography on reversed phase. Int J Vitam Nutr Res 1991;61:232–238.
3 Brubacher G, Vuilleumier JP: Vitamin C; in Curtius HC, Roth M (eds): Clinical Biochemistry Principles and Methods. New York, de Gruyter, 1974, pp 989–997.
4 Christie WW: Lipid analysis in isolation, separation, identification and structural analysis of lipids. New York, Pergamon Press, 1982, p 207.
5 Lepage G, Munoz G, Champagne J, et al: Preparative steps necessary for the accurate measurement of malondialdehyde by high-performance liquid chromatography. Analyt Biochem 1991; 197:277–283.

Dr. J. P. Allard, 9th Floor Eaton Wing, Room 223, Toronto General Hospital,
200 Elizabeth Street, Toronto, Ont. M5G 2C4 (Canada)

Galli C, Simopoulos AP, Tremoli E (eds): Fatty Acids and Lipids: Biological Aspects.
World Rev Nutr Diet. Basel, Karger, 1994, vol 75, pp 166–168

Physiological Requirements of Vitamin E as a Function of the Amount and Type of Polyunsaturated Fatty Acid

R. Muggli

Vitamins and Fine Chemicals Division, Human Nutrition Research,
F. Hoffmann-La Roche Ltd., Basel, Switzerland

High intakes of polyunsaturated fatty acids (PUFA) in animals and man lead to reduced plasma and tissue vitamin E levels and to symptoms of relative vitamin E deficiency, such as creatinuria, erythrocyte peroxidation, susceptibility of low density lipoprotein (LDL) to oxidative modification, and elevated plasma lipid peroxide markers. These deficiency symptoms can be prevented by increasing the vitamin E intake. The higher the degree of unsaturation the more pronounced or the earlier the onset of the deficiency symptoms [1, 2]. While there is clear evidence for a dependency of the vitamin E requirement on the amount and the degree of unsaturation of PUFA in the diet, the exact amount of vitamin E needed to compensate for the elevated demand caused by PUFA in the diet has not been systematically investigated in man and must be extrapolated from animal data.

Vitamin E and Lipid Peroxidation

The major function of vitamin E is the protection of the easily autoxidizable PUFA in cellular membrane lipids against oxygen free radical attack. Tocopherol does not prevent the initiation of lipid peroxidation, and at physiological levels vitamin E does not significantly react with, or destroy, hydroperoxides. Vitamin E minimizes free radical tissue damage by interrupting the chain reaction producing fatty acid hydroperoxides.

PUFA and Vitamin E Absorption

PUFA increase the requirement for vitamin E not only because cell membranes need added protection from being peroxidized, but also because diets high in polyunsaturates seem to impair the absorption of the vitamin. Weber et al. [3] studied the vitamin E uptake in rats given ^{14}C-labelled *dl*-α-tocopheryl acetate in a diet containing either 10% coconut fat or 10% linoleic acid. Less vitamin E in the tissues and more radioactivity in the feces were found in the animals fed the unsaturated fat. Diminished vitamin E absorption in the presence of large amounts of linoleic acid was also concluded in studies with fistulation of the ductus thoracicus in animals and man [4, 5].

Thus, PUFA increase the vitamin E requirements (1) by lowering the fraction of vitamin E absorbed, and (2) by causing a higher metabolic consumption. In general, studies designed to establish the tocopherol adequacy do measure the effect of both factors combined.

Vitamin E Requirement and PUFA Intake

Probably the most thorough attempt to arrive at a definite conclusion comes from Witting and Horwitt [1] who fed young tocopherol-deficient rats a variety of specially prepared fats which were, with one exception, of constant total unsaturation, but differed in composition with respect to PUFA. The time of onset of creatinuria – a marker of muscle cell membrane damage – served as a measure for the peroxidizability of the experimental diets. From these studies

Table 1. Estimated minimum requirements of vitamin E needed for compensating the elevated vitamin demand caused by some common unsaturated fatty acids

Double bonds		Fatty acid	Vitamin E requirement mg/g fatty acids	
			d-α-tocopherol	*dl*-α-tocopheryl acetate
1	Oleic acid	18:1ω9	0.09	0.13
2	Linoleic acid	18:2ω6	0.60	0.89
3	γ-Linolenic acid	18:3ω6	0.90	1.34
3	α-Linolenic acid	18:3ω3	0.90	1.34
3	Dihomo-γ-linolenic acid	20:3ω6	0.90	1.34
4	Arachidonic acid	20:4ω6	1.20	1.79
5	Eicosapentaenoic acid	20:5ω3	1.50	2.24
6	Docosahexaenoic acid	22:6ω3	1.80	2.68

it was possible to demonstrate that the relative quantities of tocopherol required to protect 1 mol of mono-, di-, tri-, tetra-, penta-, and hexaenoic acid were in the ratios 0.3:2:3:4:5:6, i.e. consistent with the relative maximum rate of autoxidation of individual PUFA in vitro which was shown to increase approximately linearly with the number of double allylic positions present in the molecule [6].

Harris and Embree [7] proposed that the tocopherol adequacy of diets and individual foods could be expressed as the ratio of the amount of vitamin E (*d*-α-tocopherol) in milligram to the amount of PUFA in gram. From animal experiments and human studies with diets rich in linoleic acid they proposed a minimal ratio of 0.6 to prevent vitamin E depletion. This critical value of 0.6 seems to be well within the general magnitude of such ratios found by other investigators [8–10] and can be used to convert the relative vitamin E requirements for individual PUFA as determined by Witting and Horwitt [1] to absolute figures (table 1). For PUFA mixtures the minimum elevated vitamin E demand can be calculated from the formula:

$$M_{vit\,E} = 0.2 \times 10^{-3} (0.3m_1 + 2m_2 + 3m_3 + 4m_4 + 5m_5 + 6m_6)$$

where $M_{vit\,E}$ = mol of *d*-α-tocopherol and m_n = mol of unsaturated fatty acid with n double bonds. As there is evidence that these estimates may not be adequate to maintain serum vitamin E levels in all circumstances it may be prudent to add a specific percentage to widen the safety margin.

References

1 Witting LA, Horwitt MK: Effect of degree of fatty acid unsaturation in tocopherol deficiency-induced creatinuria. J Nutr 1964;82:19–33.
2 Meydani SN, Shapiro AC, Meydani M, et al: Effect of age and dietary fat (fish, corn and coconut oils) on tocopherol status of C57BL/6Nia mice. Lipids 1987;22:345–350.
3 Weber F, Weiser H, Wiss O: Bedarf an Vitamin E in Äbhängigkeit von der Zufuhr an Linolsäure. Z Ernährungswiss 1964;4:1–8.
4 Hidiroglou M, Jenkins KJ, Lessard JR, et al: Effect of feeding cod liver oil on the fate of radiotocopherol in sheep. Can J Physiol Pharmacol 1970;48:751–757.
5 Blomstrand R, Forsgren L: Labelled tocopherol in man. Int Z Vitam Forsch 1968;38:328–344.
6 Cosgrove JP, Church DF, Pryor WA: The kinetics of the autoxidation of polyunsaturated fatty acids. Lipids 1987;22:299–304.
7 Harris PL, Embree ND: Quantitative consideration of the effect of polyunsaturated fatty acid content of the diet upon the requirement for vitamin E. Am J Clin Nutr 1963;13:385–392.
8 Horwitt MK: Vitamin E and lipid metabolism in man. Am J Clin Nutr 1960;8:451–461.
9 Dayton S, Hashimoto S, Rosenblum D, et al: Vitamin E status on humans during prolonged feeding of unsaturated fats. J Lab Clin Med 1965;65:739–747.
10 Lewis JS, Alfin-Slater R: An E/PUFA ratio of 0.4 maintains normal plasma tocopherol levels in growing children. Fed Proc 1969;28:758.

R. Muggli, Vitamins and Fine Chemicals Division, Human Nutrition Research,
F. Hoffmann-La Roche Ltd., Grenzacherstrasse 124, CH-4002 Basel (Switzerland)

Galli C, Simopoulos AP, Tremoli E (eds): Fatty Acids and Lipids: Biological Aspects.
World Rev Nutr Diet. Basel, Karger, 1994, vol 75, pp 169–172

A Phenolic Antioxidant Extracted from Olive Oil Inhibits Platelet Aggregation and Arachidonic Acid Metabolism in vitro

A. Petroni [a], *M. Blasevich* [a], *M. Salami* [a], *M. Servili* [b], *G.F. Montedoro* [b], *C. Galli* [a]

[a] Institute of Pharmacological Sciences, University of Milan; [b] Industrie Agrarie, University of Perugia, Italy

The great stability of olive oil toward oxidative processes is due to the presence of potent antioxidants of phenolic nature. The amounts of these compounds in olives depend upon various factors, such as climate, cultivar, stage of maturation and their concentrations in oils are in the range of 50–500 mg/kg. 2(3,4-Dihydroxyphenyl)ethanol (DHPE), oleuropeine glycoside and (*p*-hydroxyphenyl)ethanol are the most abundant phenols [1], a class of compounds that are well known for their antioxidant properties, but poorly studied for their biological activities.

This study was aimed to investigate whether DHPE affects in vitro platelet function and the production of eicosanoids, oxygenated products of arachidonic acid (AA) metabolism, which are generated through the formation of lipoperoxide intermediates. More specifically, we have evaluated in vitro the effects of DHPE on collagen-induced aggregation and thromboxane B_2 (TxB_2) formation in platelet-rich plasma (PRP), and on the accumulation of TxB_2 and of 12-hydroxyeicosatetraenoic acid (12-HETE), a major product of AA lipoxygenase, in serum, after preincubation of platelets or blood, respectively, with the compound. The data indicate that DHPE has platelet antiaggregatory activity and reduces the synthesis of both cyclo- and lipoxygenase products.

Materials and Methods

DHPE has been extracted, purified and provided to us by the group of Montedoro [1]; collagen was purchased from Gibco (Paisley, UK); all the HPLC solvents were from Carlo

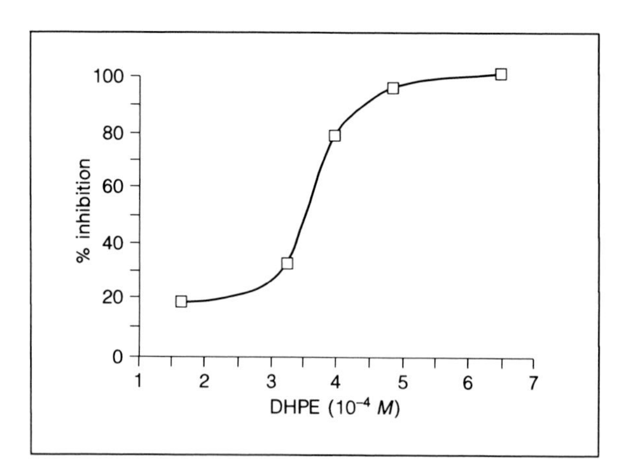

Fig. 1. Percentage inhibition of collagen-induced (4 μg/ml) aggregation of PRP (3×10^5 platelets/μl), by increasing concentrations of DHPE, added 10 min before stimulation. Shorter periods of preincubation gave lower inhibitions. Values represent the reduction of the amplitudes of the (maximal) aggregation curves obtained with the above collagen concentrations. Each value is the average of two determinations.

Erba (Milan, Italy); RP-C_{18} column (250×4.6 mm, 5-μm particles) was from Alltech Associates Inc. (Deerfield, Ill., USA). Blood was collected from human volunteers in Na citrate (10% v/v of 3.8% citrate solution), for the studies on platelet aggregation and TxB_2 production. PRP was prepared according to standardized procedures. Platelets were counted by contrast phase microscopy and PRP was brought to constant platelet count (3×10^5/μl) with platelet-poor plasma (PPP), obtained by further centrifugation of the blood remaining after removal of PRP. Aggregation of PRP was induced by collagen. The concentration of 4 μg/ml was chosen for all the experiments since it gave a maximal amplitude of aggregation. The effects of DHPE on platelet aggregation was studied after 10 min preincubation of PRP with the compound dissolved in ethanol/water 1/4, at the final ethanol concentration, in the sample, of 0.08%. Shorter periods of preincubation gave lower inhibition of PRP aggregation. The same amount of vehicle was added to control samples. The effect of DHPE on TxB_2 production was assessed by radioimmunoassay (RIA) [2] determination of TxB_2 levels in PRP, 7 min after stimulation with 4 μg/ml collagen in samples preincubated for 10 min with 6×10^{-4} *M* DHPE. Control samples were preincubated with vehicle alone. Blood was incubated with DHPE (6×10^{-4} *M*) and serum was obtained at 37 °C for 30 min. Under these conditions marked accumulation of eicosanoids occurs [3], due to maximal stimulation of blood cells. Serum was used for TxB_2 measurements by RIA [3] or extracted with ethylacetate and injected in HPLC for the evaluation of 12-HETE, made quantitative by the use of 13-hydroxyoctadecadienoic acid (13-HODE) as internal standard.

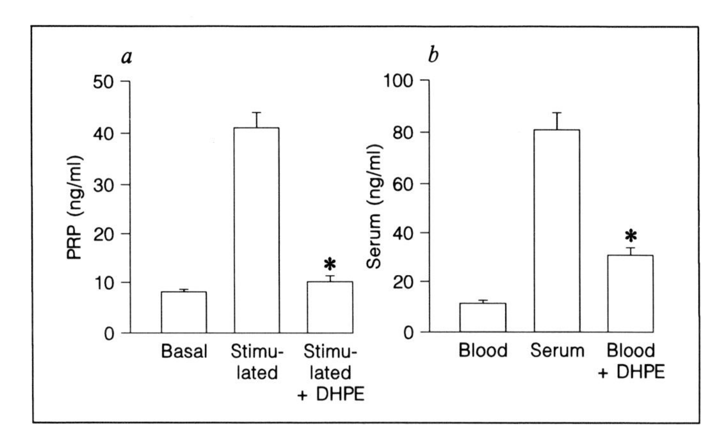

Fig. 2. Levels (ng/ml) of TxB_2, measured by RIA in PRP and in serum (*b*). Stimulation of PRP with 4 μg/ml collagen for 7 min resulted in marked elevation of TxB_2 over basal levels (*a*), and the increment was almost completely prevented by preincubation with 6×10^{-4} *M* DHPE, a concentration which completely inhibited PRP aggregation. Preincubation of blood with the same concentration of DHPE for 10 min markedly reduced the accumulation of TxB_2, occurring during serum formation (37 °C for 30 min). Values are the average ± SE of three determinations. * Significantly different ($p < 0.02$, student's t-test) vs stimulated or serum.

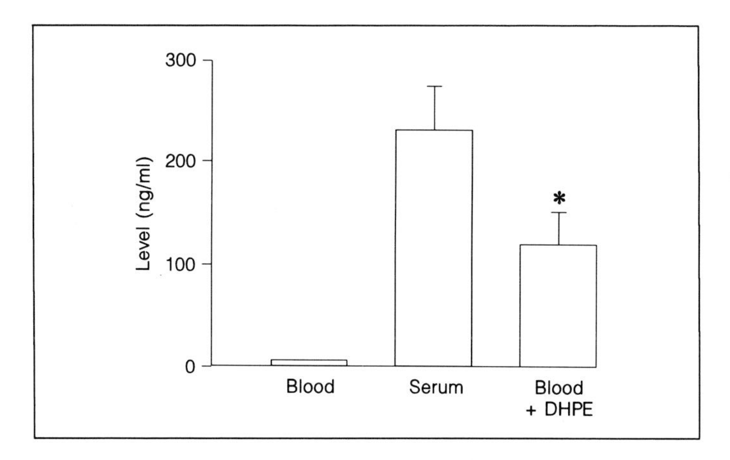

Fig. 3. Levels (ng/ml) of 12-HETE, a major lipoxygenase product, in serum obtained in the absence or in the presence of 6×10^{-4} *M* DHPE. This compound reduced the accumulation of TxB_2 in serum by about 50%. * Significantly different ($p < 0.05$, student's t-test) vs serum.

Results

PRP aggregation, induced by collagen at the concentration of 4 μg/ml and measured 7 min after stimulation, was inhibited in a dose-dependent manner by DHPE incubated for 10 min before collagen (fig. 1). The IC_{50}, i.e. the concentration giving 50% inhibition, was 3×10^{-4} *M*. The levels of TxB_2, a very potent aggregating agent, measured by RIA in PRP prior and after collagen stimulation, and in serum versus blood are shown in figure 2. Preincubation of PRP with DHPE, at the concentration which completely inhibited PRP aggregation, i.e. 6×10^{-4} *M*, before collagen stimulation, resulted in complete inhibition of TxB_2 production, and, when preincubated with blood, in about 70% reduction of TxB_2 levels. The levels of 12-HETE measured in serum obtained from blood without or in the presence of DHPE, are shown in figure 3. DHPE inhibited by 50% the accumulation of 12-HETE in serum.

Discussion

The effects of DHPE on platelet function appear very interesting, when compared for instance with those of other compounds such as aspirin or vitamin E. In fact, although aspirin is more potent than DHPE, with an IC_{50} on collagen-induced PRP aggregation of the order of 10^{-5} *M*, it has no effect on lipoxygenase metabolites. The most important natural antioxidant, vitamin E has no effect on platelet aggregation even at the concentration 10^{-3} *M* and for longer time of incubation [4]. The combined effects of DHPE, and possibly of other phenolic compounds in olives, on platelets and on lipid peroxidation, could contribute to the favorable effects attributed to olive oil, as part of the Mediterranean diet, toward cardiovascular diseases.

References

1 Montedoro GF, Servili M, Baldioli M, et al: Simple and hydrolyzable phenolic compounds in virgin olive oil. Their extraction, separation, and quantitative and semiquantitative evaluation by HPLC. J Agric Food Chem 1992;40:1571–1576.

2 Granstrom E, Kindahl H, Samuelsson B: Radioimmunoassay for thromboxane B_2. Analyt Lett 1976;9:611–627.

3 Patrono C, Ciabattoni G, Pinca E, et al: Low dose aspirin and inhibition of thromboxane B_2 production in healthy subjects. Thromb Res 1980;17:317–327.

4 Agradi E, Petroni A, Socini A, et al: In vitro effects of synthetic antioxidants and vitamin E on arachidonic acid metabolism and thromboxane formation in human platelets and in platelet aggregation. Prostaglandins 1981;22:255–266.

Claudio Galli, MD, Institute of Pharmacological Sciences, University of Milan,
Via Balzaretti, 9, I–20133 Milano (Italy)

Galli C, Simopoulos AP, Tremoli E (eds): Fatty Acids and Lipids: Biological Aspects.
World Rev Nutr Diet. Basel, Karger, 1994, vol 75, pp 173–174

Summary Statement: Isomeric Fatty Acids

Ronald P. Mensink

The session was co-chaired by *R. P. Mensink* and *P. J. Nestel,* and presentations were made by Drs. *Mensink, Nestel, A. C. van Houwelingen, M. Sugano, K. W. J. Wahle,* and *D. A. Wood.*

During this session various physiological effects of *trans* monounsaturated fatty acids (trans-C18:1) were discussed. *Trans*-C18:1 may stimulate platelet aggregation. Furthermore, *trans*-C18:1 has an inhibiting effect on elongation and desaturation of the EFA, LA and LNA. This effect can be overcome by increasing EFA availability, which can easily be achieved by dietary means. *Trans*-C18:1, however, is already present in the developing fetus, and phospholipids isolated from fetal tissue contain very low amounts of LA, whereas the LNA content is hardly measurable. Under this particular condition, the inhibiting effect of *trans*-C18:1 on EFA desaturation and elongation may have a considerable influence on the availability of the longer chain EFA derivatives, which are instrumental for fetal development. Since fetal *trans*-C18:1 originate from the mother, these data suggest that the maternal diet should be as low as possible in *trans* fatty acids.

Trans-C18:1 also have a negative effect on the serum lipoprotein profile. All well-controlled studies performed so far have shown a LDL cholesterol-raising effect of *trans*-C18:1 relative to cis-C18:1 (oleic acid). In addition, most studies suggest an HDL-lowering effect and Lp(a)-increasing effect of *trans*-C18:1. Finally, a case-control study suggested that *trans*-isomers of LA were associated with sudden cardiac death. It should be mentioned, however, that no such relation was found for *trans*-C18:1.

In conclusion, several studies suggest that *trans*-C18:1 may have adverse effects on health. The question arises, however, if products high in *trans* are indeed not beneficial, are there alternatives? Food technologists do need fats with a melting point range between about 30 and 40 °C to make products that should also be palatable for the consumer. It should be their challenge to compose fats and oils low, or even free, in *trans* fatty acids, which meet these requirements without adversely affecting human health. In the meantime, studies should continue so as to obtain reliable data on the intake of *trans* fatty acids and to describe their physiological and metabolic effects in more detail.

Galli C, Simopoulos AP, Tremoli E (eds): Fatty Acids and Lipids: Biological Aspects.
World Rev Nutr Diet. Basel, Karger, 1994, vol 75, pp 175–178

Trans Fatty Acids in Early Human Development

Adriana C. v. Houwelingen, Gerard Hornstra

Department of Human Biology, Limburg University, Maastricht, The Netherlands

Unsaturated fatty acids have one or more double bonds. The configuration of these double bonds can be either *cis* or *trans*. Most of the unsaturated fatty acids found in nature have the *cis* configuration, whereas *trans* fatty acids originate primarily from technically hydrogenated fats. However, they are also present in fats from ruminants (for instance milk fat and butter) and in the meat of these animals. *Trans* fatty acids account for 10% of polyenoic and for about 90% of monoenoic fatty acids. *Trans* fatty acids are markedly different from their respective *cis* isomers in melting point and interactions with enzymes and metabolism.

A number of untoward biological effects of *trans* fatty acid consumption have been observed. Studies in human fibroblasts in vitro have demonstrated that *trans* fatty acids impair the microsomal desaturation and chain elongation of the essential fatty acids linoleic and α-linolenic acids to their long chain polyunsaturated metabolites [1]. These metabolites are of great importance during perinatal development. Mensink and Katan [2] concluded that the effect of *trans* fatty acids on the lipoprotein profile is at least as unfavorable as that of the cholesterol-raising saturated fatty acids which means that high density lipoprotein (HDL) cholesterol levels fall, and low density lipoprotein (LDL) cholesterol levels increase. Two recent papers [3, 4] reported that dietary *trans* fatty acids increase serum lipoprotein (a) (Lp(a)). Therefore, *trans* fatty acid exposure in early life may lead to an undesirable increase in these atherogenic risk factors already in infancy. In addition, Koletzko [5] observed an inverse relationship between the amount of 18:1 *trans* fatty acids in cholesterol esters in plasma of premature infants and their birthweight, which might be an indication for impaired early growth. In the paper of Opstvedt and Pettersen [6], intrauterine growth retardation was suggested by a trend to lower birthweights in piglets born to sows fed partially hydrogenated fish oil or soybean oil. Yu et al. [7] and Hill et al. [8] showed that *trans* unsaturated fatty acids fed to newborn rats and mice impaired postnatal weight gain. Therefore, potential adverse effects of *trans* isomers at the present level of dietary intake are conceivable.

Trans fatty acids are consumed in rather large amounts in the industrialized countries, on an average 2–12 g/day, which equals about 2.5% of total energy or 6–8% of the total fat intake [9]. The consumption of *trans* fatty acids seems to have decreased over the last years due to the progress in food technology and changes in nutritional habits.

Materials and Methods

We have determined the dietary intake of *trans* fatty acids of Dutch pregnant women by analyzing the fatty acid composition of duplicate portions of their food collected during 4 days. Moreover, we investigated the *trans* fatty acid content of phospholipids isolated from: (a) human fetal tissue, collected after elective abortions performed at a gestational age varying from 5 to 15 weeks ($n = 38$), and (b) umbilical venous plasma ($n = 28$) and umbilical vessel walls ($n = 37$) from full-term neonates. With almost each fetal sample a matched maternal venous blood sample could be analyzed also.

Results and Discussion

Dutch pregnant women consumed on average 3 g *trans* fatty acids per day. The amounts of most fatty acids in fetal tissue showed an increase with gestational age [10]. The amounts of *trans* 18:1 isomer, however, did not increase during pregnancy. Hence, the relative amounts decreased during early gestation. Even at a gestational age of 5 weeks, *trans* isomers were detected in fetal tissues. A highly significant correlation between relative amounts of 18:1 *trans* in maternal plasma (= y) and fetal tissue (= x) was observed ($y = 0.15 + 0.38x$; $r = 0.63$; $n = 38$; $p = 0.0001$) (fig. 1). Also at delivery a correlation between maternal (= y) and fetal umbilical (= x) plasma values was observed ($y = 0.13 + 1.2x$; $r = 0.71$; $p = 0.001$; $n = 28$). From these results we may conclude that the degree of intrauterine *trans* fatty acid exposure is related to maternal values of *trans* fatty acids.

To investigate the possible influence of *trans* unsaturated fatty acids on the metabolism of essential fatty acids, the amounts of 18:1 *trans* were correlated with some ratios known to reflect essential fatty acid chain elongation and desaturation. To exclude the influence of the course during gestation we corrected for gestational age. First the ω6 metabolic index (ratio of ω6

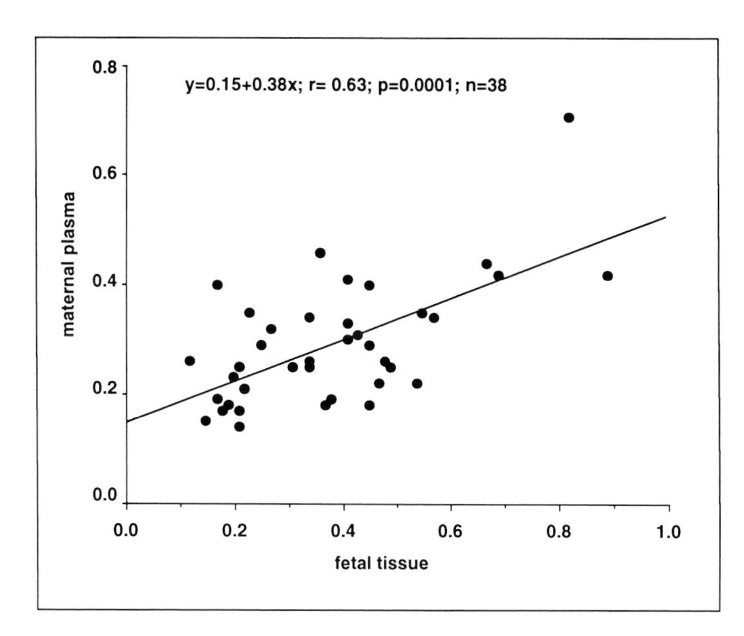

Fig. 1. Correlation between maternal plasma and fetal tissue 18:1ω9 *trans* (%) in the first trimester of gestation (5–15 weeks).

LCPUFA/18:2ω6), which reflects the conversion of linoleic acid into its longer chain, more unsaturated derivatives: the ω6 long chain polyunsaturates (ω6 LCPUFA). In fetal tissues (gestational age 5–15 weeks) an inverse relation with the 18:1ω9 *trans* isomer was found (partial $r = -0.75$; $p = 0.0003$), which indicates the lower the amounts of ω6 elongation and desaturation products, the higher the amounts of *trans* fatty acids.

The docosahexaenoic acid (DHA) deficiency index (ratio of 22:5ω6/ 22:4ω6) showed a positive relation with the *trans* isomers (partial $r = -0.37$; $p = 0.0086$). This might point to an inhibitory effect on the ω3 metabolism also or to a stimulation of the Δ4-desaturase activity in the ω6 family. A significant inverse relation between the 18:1ω9 *trans* and the ω6 metabolic index (partial $r = -0.30$; $p = 0.048$) was also found in the umbilical arterial walls of full-term neonates. This confirms the earlier suggestion that the lower the amounts of ω6 metabolic products, the higher the amounts of *trans* unsaturated fatty acids present.

The DHA deficiency index showed a positive relation with the *trans* isomers also (partial $r = -0.32$; $p = 0.034$). In combination with the observed inverse relation between DHA and the *trans* unsaturated fatty acids, this might point to an inhibitory effect on ω3 metabolism rather than to a stimulation of the Δ4 desaturation activity in the ω6 family.

Finally, the relation was investigated between 18:1 *trans* in umbilical arterial vessel walls and some clinical variables. For birthweight and head circumference, significant correlations were observed with 18:1ω9 *trans* ($r = -0.35$; $p = 0.033$, and $r = -0.37$; $p = 0.028$, respectively).

Trans fatty acids in the fetus/neonates tend to increase the need for essential fatty acids, which are of great importance during perinatal development. Our studies both in fetal tissue and in umbilical vessel walls demonstrated that *trans* isomers of fatty acids interfere with essential fatty acid elongation and desaturation. Therefore a further deterioration of the already marginal perinatal long chain polyenes status by *trans* fatty acids is feasible, which is likely to have adverse effects on human development in the perinatal period.

References

1 Rosenthal MD, Doloresco MA: The effects of *trans* fatty acids on fatty acyl Δ5 desaturation by human skin fibroblasts. Lipids 1984;19:869–874.

2 Mensink RP, Katan MB: Effect of dietary *trans* fatty acids on high-density and low-density lipoprotein cholesterol levels in healthy subjects. N Engl J Med 1990;323:439–445.

3 Nestel PJ, Noakes M, Belling GB, et al: Plasma lipoprotein lipid and Lp(a) changes with substitution of elaidic acid for oleic acid in the diet. J Lipid Res 1992;33:1029–1036.

4 Mensink RP, Zock PL, Katan MB, Hornstra G: Effect of dietary *cis*- and *trans*-fatty acids on serum lipoprotein (a) levels in humans. J Lipid Res 1992;33:1493–1501.

5 Koletzko B: *Trans* fatty acids may impair biosynthesis of long chain polyunsaturates and growth in men. Acta Paediatr 1992;81:302–306.

6 Opstvedt J, Pettersen J, Mork SJ: *Trans* fatty acids. 1. Growth fertility, organ weight and nerve histology and conduction velocity in sows and offspring. Lipids 1988;23:713–719.

7 Yu PH, Mai J, Kinsella JE: The effect of dietary *trans* methyl octadienoate acid on composition and fatty acids of the heart. Am J Clin Nutr 1980;33:598–605.

8 Hill EG, Johnson SB, Leason LD, et al: Perturbation of the metabolism of essential fatty acids by dietary partially hydrogenated vegetable oil. Proc Natl Acad Sci USA 1982;79:953–957.

9 Mensink RP, Katan MB: *Trans* monounsaturated fatty acids in nutrition and their impact on serum lipoprotein levels in man. Prog Lipid Res 1993;32:111–122.

10 Houwelingen ACv, Puls J, Hornstra G: Essential fatty acid status during early human development. Early Hum Dev 1992;31:97–111.

Dr. A.C.v. Houwelingen, Department of Human Biology, Limburg University,
PO Box 616, NL-6200 MD Maastricht (The Netherlands)

Galli C, Simopoulos AP, Tremoli E (eds): Fatty Acids and Lipids: Biological Aspects.
World Rev Nutr Diet. Basel, Karger, 1994, vol 75, pp 179–182

Trans Fatty Acids: Effects on Eicosanoid Production

Michihiro Sugano[a], *Tamiho Koga*[b], *Takako Yamato*[b],
Michiko Nonaka[a], *Jong Yan Gu*[a]

[a] Laboratory of Food Science, Kyushu University School of Agriculture, Fukuoka;
[b] Laboratory of Food Processing, Nakamura Gakuen College, Fukuoka, Japan

Because of their artificial nature, *trans* fatty acids, exclusive of *trans* octadecaenoic acid, have long been regarded as 'bad' fatty acids. In most studies, however, comparison has been made between nonhydrogenated and hydrogenated fats, and it is difficult to attribute the observed effects to *trans* fatty acids alone, since there is a simultaneous reduction of polyunsaturated fatty acids and an increase in saturated fatty acids during partial hydrogenation of vegetable oils. The comparison, therefore, must be done under a dietary regimen where the difference in the geometry of octadecaenoic acid is the sole variable. It is of course also necessary to supply an adequate amount of linoleic acid (LA).

Effect on LA Desaturation and Prostaglandin Production

In animal studies *trans* octadecaenoic acid appears to interfere with the desaturation of LA to arachidonic acid (AA) as estimated from the reduction of the ratio of AA/LA in tissue phospholipids, even though the experimental diets satisfy the above criteria [1–5].

Nevertheless, *trans* fatty acids do not appear to affect the production of prostaglandins such as PGE_1, PGE_2, PGI_2 and TXA_2 [1–4]. In addition, in rat kidney the ratio of vasodilating PGE_2/vasoconstricting $PGF_{2\alpha}$ increased significantly in stroke-prone SHRs given *trans* fat compared to those given *cis* fat, suggesting a preventive function against hypertension [5]. Thus, the available

Table 1. Effects of *trans* fat on plasma PGE_2 concentration and LTC_4 production in rat spleen

Groups	n	PGE_2 pg/ml plasma	LTC_4 ng/g spleen
Casein diets			
Cis fat	8	7.26 ± 2.74	43.5 ± 5.8*
Trans fat	8	7.49 ± 2.31	28.1 ± 3.0
Soybean protein diets			
Cis fat	8	15.3 ± 1.6	36.0 ± 2.9
Trans fat	8	13.7 ± 3.5	39.0 ± 4.2

Rats were fed purified diets containing 10% fat with either 31.8% of *trans* 18:1 and 33.4% *cis* 18:1 or 65.4% *cis* 18:1 for 3 weeks.
* Significantly different from the corresponding *trans* fat group at $p < 0.05$.

data suggest that *trans* octadecaenoic acid does not appear to affect the production of eicosanoids unless the supply of LA is at least nutritionally inadequate [6].

Effects on Leukotriene Production

The production of hydroxy fatty acids from AA is also not influenced by *trans* fatty acids [3]. We have recently studied in rats the interaction of *trans* fatty acids with dietary protein, either casein or soybean protein. The fatty acid composition data indicated that the effect of *trans* fatty acids on the conversion of LA to AA in the spleen depends on the protein source. *Trans* fatty acids significantly reduced the splenic production of leukotriene C_4 (LTC_4) on a unit weight base when rats were fed casein, but not soybean protein (table 1). However, no effect of *trans* fatty acids was found on the concentration of plasma PGE_2.

The extent of incorporation of *trans* fatty acids into liver glycerophospholipids was more marked with soybean protein than with casein. More *trans* fatty acids were detected in phosphatidylinositol than in phosphatidylcholine or phosphatidylethanolamine. In contrast, practically no *trans* fatty acid was incorporated into cardiolipin.

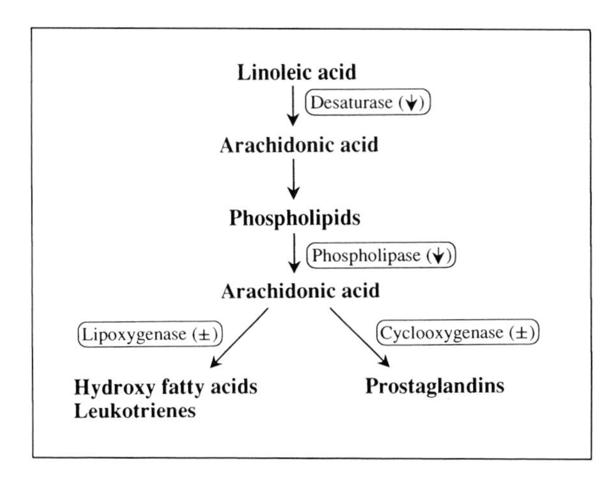

Fig. 1. Possible effects of *trans* octadecaenoic acid on AA casade.

Conclusion

Changes in the metabolism of polyunsaturated fatty acids induced by dietary *trans* monoene fatty acids are summarized in figure 1. *Trans* fatty acids reduce the formation of AA from LA and hence, lower the pool size of AA in phospholipids even when the supply of LA is adequate. However, the amount of the substrate fatty acid required for the production of eicosanoids is practically marginal, and *trans* fatty acids do not modify the enzyme activity involved in the production of prostaglandins and leukotrienes. Consequently, dietary *trans* fatty acids do not necessarily affect the level of eicosanoids. Thus, the net result of feeding *trans* fatty acids is that this unnatural fatty acid may not modify the metabolic control system in the animals unless the supply of LA is unreasonably low.

References

1 Watanabe M, Koga T, Sugano M: Influence of dietary *cis* and *trans* fats on 1,2-dimethylhydrazine-induced colon tumorigenesis and fecal steroid excretion in Fischer 344 rats. Am J Clin Nutr 1985; 42:475–484.
2 Sugano M, Watanabe M, Yoshida K: Influence of dietary *cis* and *trans* fats on DMH-induced colon tumors, steroid excretion, and eicosanoid production in rats prone to colon cancer. Nutr Cancer 1989;12:177–187.
3 Zevenbergen JL, Haddeman E: Lack of effects of *trans* fatty acids on eicosanoid biosynthesis with adequate intake of linoleic acid. Lipids 1989;24:555–563.
4 Høy CE, Hølmer G: Influence of dietary linoleic acid and *trans* fatty acids on the fatty acid profile of cardiolipins in rats. Lipids 1990;25:455–459.

5 Chiang MT, Otomo MI, Itoh H, et al: Effect of *trans* fatty acids on plasma lipids, platelet function and systolic blood pressure in stroke-prone spontaneously hypertensive rats. Lipids 1991;26:46–52.

6 Kinsella JE, Bruckner G, May J, et al: Metabolism of *trans* fatty acids with emphasis on the effects of trans, trans-octadecadienoate on liver lipid composition, essential fatty acid, and prostaglandins: An overview. Am J Clin Nutr 1981;34:2307–2318.

Michihiro Sugano, PhD, Laboratory of Food Science, Department of Food Science and Technology 46–09, Kyushu University School of Agriculture, Higashi-ku, Fukuoka-shi, Fukuoka 812 (Japan)

Galli C, Simopoulos AP, Tremoli E (eds): Fatty Acids and Lipids: Biological Aspects.
World Rev Nutr Diet. Basel, Karger, 1994, vol 75, pp 183–186

Effect of *Trans* Fatty Acids on Platelet Function

Klaus W. J. Wahle, Lesley I. L. Peacock, Alison Taylor, Charles R. A. Earl

Rowett Research Institute, Bucksburn, Aberdeen, UK

Sources of Isomeric Fatty Acids in the Diet and Levels of Consumption

The major sources of isomeric *cis* and *trans* fatty acids in the human diet are the various margarines, shortenings and frying oils commercially derived from edible vegetable oils by a process of chemical 'hardening' or hydrogenation. The extent of this hydrogenation depends on the desired stability and physical characteristics of the end product as in stick or tub margarines or in salad dressings [1]. Hydrogenation of oils reduces the natural *cis* double bonds in the fatty acid carbon chains thereby decreasing the proportions of essential fatty acids (EFA). The process is relatively imprecise and a variety of positional *cis* and *trans* isomers are formed which can comprise up to 60% of the total fatty acids of the product. Positional isomers of monoenoic 18:1 are formed when vegetable oils are hydrogenated and *trans*-9–11 18:1 predominate; up to 10% of fatty acids in hardened oils may be dienoic 18:2. The complexity of the isomeric mix increases when fish oils, with their variety of long-chain polyunsaturated fatty acids (PUFA) (18–22 carbons), are hydrogenated [2–4]. Surprisingly, little is known about the metabolic effects of *cis* and *trans* isomers of 20 and 22 carbon fatty acids.

Ruminant fats also contain isomeric fatty acids due to the biohydrogenation of feed PUFA in the rumen but these isomers comprise only 2-5% of total fatty acids, with *trans*-11 18:1 predominating, and they generally only comprise about 15–20% of total *trans* intake in man [3–5]. Average per capita intakes of *trans* fatty acids are 7–8 g/day in the UK and USA, but these averages may be misleading insofar that certain subpopulations may consume 27 g/day or more [4, 6]. A recent report estimates *trans* consumption in the USA to range from 11.1 to 27.6 g/day [7]. Isomeric fatty acids are found in most animal and human tissues in concentrations usually reflecting their levels in the diet. They are present in all lipid classes and levels in phospholipids, although lower than neutral lipids, are considered as potentially the most important with regard to their possible deleterious effects on health (see below).

Isomeric Fatty Acids and Cardiovascular Disease

Dietary isomeric fatty acids, particularly *trans* isomers, have been implicated in the aetiology of cardiovascular disease (CVD) but not without controversy [3, 4, 6, 8]. Observed detrimental effects of isomeric fatty acids in earlier studies were criticized because of attendant EFA deficiencies. However, recent epidemiological and dietary studies strongly indicate that isomeric *trans* fatty acids derived from hydrogenated vegetable oils are more atherogenic than those in ruminant fats [5] and have detrimental effects on indices of CVD risk such as LDL and HDL cholesterol and lipoprotein (a) concentrations [9–12] despite adequate EFA intakes. Furthermore, the *trans* isomer content of adipose tissue was higher in men who had died from CVD than in those who died from other causes [13] and a positive relationship was apparent between the linoleic/*trans* fatty acid ratios in adipose tissue and the standardized mortality for CVD in two Scottish cities [14]. No biochemical or metabolic mechanisms have been proposed to explain these detrimental effects of isomeric fatty acids on the cardiovascular system.

Platelet Function

Thrombosis is the major cause of death in CVD and the sensitivity of platelets to aggregation by various agonists, which can be modified by dietary lipids, is of primary importance in predisposing to thrombus formation [15]. Surprisingly, few researchers have investigated the possible detrimental effects of isomeric fatty acids on platelet function and thrombogenesis and the results obtained have often been contradictory, due possibly to species differences, different platelet preparations and/or to the optical method used to determine aggregation.

Cis isomers of mono- and diunsaturated fatty acids inhibited both thrombin- and collagen-induced aggregation of washed human platelets but the corresponding *trans* isomers had no effect on thrombin-induced aggregation: collagen-stimulated aggregation was unfortunately not determined [16]. Neither *cis* nor *trans* mono- and diunsaturated isomers reduced thrombin-induced aggregation of rabbit platelets but they all inhibited aggregation stimulated by collagen, arachidonic acid (20:4ω6, AA) or thromboxane mimetics to a similar extent [17]. A number of *cis* monoenoic isomers (9, 11, 12, 13) inhibited both thrombin ($>80\%$) and collagen ($>70\%$) stimulated aggregation in porcine platelets, which resemble human platelets in structure, composition and reactivity to agonists [18]. However, the corresponding *trans* isomers were much less effective (20–60% inhibition) when collagen was the agonist and they actually augmented (10–30%) aggregation induced by thrombin. Both *cis* and *trans* isomers significantly inhibited thromboxane formation during aggregation with collagen or thrombin [18]. Unlike various *cis* isomers, *trans*-9 18:1 did not inhibit ADP-induced aggregation in bovine platelets [19]. Inhibition of these receptor-mediated platelet responses by isomers was elicited by the free fatty acids intercalating into the platelet membrane because incorporation into membrane complex lipids was low [16] or virtually absent [18]. The effects of the *cis* isomers on platelet function were possibly due to increased membrane fluidity which they are reported to elicit but the *trans* isomers did not influence fluidity [16, 17, 19]. This suggests the involvement of mechanism(s) other than perturbation of membrane fluidity for the *trans* isomer effects since they inhibited aggregation and TXA_2 without affecting fluidity. Contrasting observations regarding the effect of different isomeric fatty acids on platelet aggregation may relate to a lack of sensitivity of the optical method when lipids are used in aggregation studies. The impedence method used by Wahle et al. [18] is more sensitive and should be the method of choice when lipids are studied.

Isomeric fatty acids did not appear to alter agonist-membrane receptor interactions [16] or phospholipase A_2 and C activities [17, 20] in platelets from different species. In porcine platelets the incorporation or reincorporation of AA into phosphatidylinositol (PI) and phosphatidylserine (PS), membrane phospholipids which are of pivotal importance in membrane signal transduction mechanisms, is markedly reduced by *cis* and *trans* 13C-18:1 in both the resting and stimulated state with the *trans* isomer being the most effective in PI [20]. Reduced AA availability in platelet membrane PI and PS could in part explain the reduced thromboxane formation elicited by isomeric fatty acids and also indicates a specific influence of *trans* isomers on the signal transduction pathways in platelets.

Conclusions

Epidemiological, dietary and biochemical studies are increasingly suggesting that isomeric fatty acids, particularly *trans* isomers, may play a role in the aetiology of CVD. The possibility that these isomers may exert their deleterious effects by modulating EFA metabolism and/or signal transduction pathways in platelets and possibly other cells of the blood-vascular system is cause for some concern and urgently requires further investigation.

References

1 Dutton HJ: Hydrogenation of fats and its significance; in Emken EA, Dutton HJ (eds): Geometric and Positional Fatty Acid Isomers. Champaign, American Oil Chemists' Society, 1979, pp 99–129.
2 Gurr MI: *Trans* fatty acids. Int Dairy Fed Bull 1983;166:5–18.
3 Senti FR (ed): FASEB Report. Health Aspects of Dietary *Trans* Fatty Acids, August 1985. Bethesda, Federation of American Societies for Experimental Biology, 1988.
4 BNF Report: Report of the British Nutrition Foundation's Task Force on Trans Fatty Acids. London, British Nutrition Foundation, 1987.
5 Wahle KWJ, James WPT: Isomeric fatty acids and human health. Eur J Clin Nutr 1993;47:828–839.
6 Enig MG, Subodh A, Keeney M, Samugna J: Isomeric *trans* fatty acids in the US diet. J Am Coll Nutr 1990;9:471–486.
7 Beare-Rogers IL: *Trans-* and positional isomers of common fatty acids. Adv Nutr Res 1988;5:171–200.
8 Willett WC, Stampfer MJ, Manson JE, Colditz GA, Speizer FE, Rosner BA, Sampson LA, Hennekens CH: Intake of *trans* fatty acids and risk of coronary heart disease among women. Lancet 1993;341:581–585.
9 Mensink RPM, Katan MB: Effect of dietary *trans* fatty acids on high-density and low-density lipoprotein cholesterol levels in healthy subjects. N Engl J Med 1990;323;439–445.
10 Mensink RPM, Zock PL, Katan MB, Hornstra G: Effect of dietary *cis* and *trans* fatty acids on serum lipoprotein (a) levels in humans. J Lipid Res 1992;33:1493–1501.
11 Nestel P, Noakes M, Belling B, McArthur R, Clifton P, Janus E, Abbey M: Plasma lipoprotein lipid and Lp(a) changes with substitution of elaidic acid for oleic acid in the diet. J Lipid Res 1992; 33:1029–1036.
12 Wahle KWJ, Mutalib SM, Whiting P, Cummings A, Broom J: The role of different dietary fats and antioxidants in regulating plasma Lp(a) concentrations in man. Abstr 1st Int Congr Int Soc Study of Fatty Acids and Lipids (ISSFAL), Lugano 1993, pp 1–137.
13 Thomas LH, Winter JA, Scott RG: Concentration of 18:1 and 16:1 transunsaturated fatty acids in the adipose body tissue of descendents dying of ischaemic heart disease compared with controls: Analysis by gas liquid chromatography. J Epidemiol Commun Health 1983;37:16–21.
14 Wahle KWJ, McIntosh G, Duncan WRH, James WPT: Concentrations of linoleic acid in adipose tissue differ with age in women but not men. Eur J Clin Nutr 1991;45;195–202.
15 Lagarde M, Gualde N, Rigaud M: Metabolic interactions between eicosanoids in blood and vascular disease. Biochem J 1989;257:313–320.
16 MacIntyre DE, Hoover RL, Smith M, Steer M, Lynch C, Karnovsky MJ, Salzman EW: Inhibition of platelet function by *cis*-unsaturated fatty acids. Blood 1984;63:848–857.
17 Sato T, Nakao K, Hashizume T, Fujii T: Inhibition of platelet aggregation by unsaturated fatty acids through interference with thromboxane-mediated process. Biochim Biophys Acta 1987;931: 157–164.
18 Wahle KWJ, Peacock LIL, Earl CRA: Effect of isomeric *cis* and *trans* monoenoic fatty acids on porcine platelet aggregation and thromboxane synthesis. Biochim Biophys Acta 1994(submitted).
19 Kitigawa S, Endo J, Kametani F: Effects of long-chain *cis* unsaturated fatty acids and their alcohol analogs on aggregation of bovine platelets and their relation with membrane fluidity changes. Biochim Biophys Acta 1985;818:391–397.
20 Wahle KWJ, Peacock LIL, Earl CRA: Effect of isomeric *cis* and *trans* monoenoic fatty acids on arachidonic acid incorporation into membrane phospholipids of porcine platelets. Biochim Biophys Acta 1994(submitted).

Klaus W. J. Wahle, MD, Rowett Research Institute, Bucksburn, Aberdeen AB2 9SB (UK)

Galli C, Simopoulos AP, Tremoli E (eds): Fatty Acids and Lipids: Biological Aspects.
World Rev Nutr Diet. Basel, Karger, 1994, vol 75, pp 187–189

Effect of *Trans* Fatty Acids on Serum Lipoprotein Levels in Man

P.J. Nestel

CSIRO Division of Human Nutrition, Adelaide, S.A., Australia

Commercial hydrogenation of polyunsaturated oils, as well as hydrogenation of linoleic (LA) and α-linolenic (LNA) acids in the rumen of sheep and cattle, lead to the production of *trans* fatty acids. Concern has been expressed over the possible consequences to health brought about by the increased consumption of *trans* fatty acids in recent years. The average amounts currently eaten in countries where hydrogenated fats occur widely in foods have been estimated as at least 4% of total energy and possibly 2–3 times that: USA, 13.3 ± 1.1 g/day [1]; 8.1 g/day [2]; 4 ± 1.9 g/day [3]; The Netherlands, 17 g/day; Australia, 3–4 g/day.

Reports on the effects of *trans* fatty acids have been inconsistent. Mattson et al. [4] found similar plasma cholesterol concentrations when elaidic acid was partly substituted for oleic acid; Mensink and Katan [5] showed that at higher intakes of elaidic acid, low density lipoprotein (LDL) cholesterol rose and high density lipoprotein (HDL) cholesterol fell, and Laine et al. [6] concluded that, provided the LA content of the diet was high enough, partial hydrogenation of LA and oleic acid would not raise the plasma cholesterol level. This has been confirmed by Nestel et al. [7]. The implications for public health and for the edible oil industry are considerable. We have therefore conducted a controlled double-blind study in which we tested a more moderate amount of *trans* fatty acid and in which the elaidic acid-rich diet was compared with an oleic acid-rich diet, as well as with two additional diets enriched with saturated fatty acids [8]. The effect of additional dietary *trans* fatty acids (7% energy) on plasma lipids was assessed in a double-blind comparison of four separate diets: (1) enriched with butter fat (lauric-myristic-palmitic); (2) oleic acid-rich; (3) elaidic acid-rich, and (4) palmitic acid-rich. In 27 mildly hypercholesterolemic men, total and LDL cholesterol were significantly lower during the 3-

Table 1. Plasma and lipoprotein lipids (mg/dl)

	Habitual	Oleic	Elaidic	Palmitic
Plasma cholesterol	228 ± 29	215 ± 28	229 ± 29	226 ± 28
Difference vs. oleic	13 ± 17*		14 ± 14*	11 ± 14*
Plasma triglyceride	139 ± 50	135 ± 42	142 ± 48	128 ± 44
LDL cholesterol	163 ± 25	151 ± 26	165 ± 29	161 ± 26
Difference vs. oleic	12 ± 14*		14 ± 12*	10 ± 10*
HDL cholesterol	38 ± 6	38 ± 6	38 ± 7	42 ± 6
Difference vs. palmitic	4 ± 4*	4 ± 4*	4 ± 3*	

* $p<0.001$.

Table 2. Do *trans* fatty acids (TFA) raise LDL-like saturates?

Study	Ref.
In 748 men, TFA intake directly correlated with LDL cholesterol (r = 0.09)	9
Hard margarine (29% TFA) gave similar LDL cholesterol as US habitual diet but *lower than butter fat* and higher than soft margarine	10
TFA (7% en) gave similar LDL cholesterol as habitual and palm oil-rich diets but higher than oleic-rich diet	8
TFA (4% en) gave lower LDL than habitual but higher than 0.4% TFA, corn oil diet	11

week oleic acid-rich diet, and were similar during the other three diets (table 1). We conclude that 3 weeks' consumption of *trans* fatty acid (mainly elaidic) at about 7% energy (probably twice the Australian average) results in LDL cholesterol levels that do not differ from those seen with diets enriched with palmitic acid or butter fat and are higher than when oleic acid is substituted for elaidic acid. It seems probable from these and other studies that *trans* fatty acids behave like saturated fatty acids in raising LDL cholesterol (table 2).

Another important observation from this study is that elaidic acid-rich diets significantly elevate plasma lipoprotein (a) (Lp(a)) levels compared to all the other diets studied, even those that produced very similar LDL cholesterol levels [8]. Plasma Lp(a) is a significant risk factor for coronary artery disease. An unresolved important question is whether HDL cholesterol is lowered. Whereas this has been shown by the Netherlands' group in two experiments [5], this was not the case in two other studies using slightly less elaidic acid [8, 10].

Dietary *trans* fatty acids are otherwise considered safe. In this study we also excluded the possibility that elaidic acid may render LDL more oxidizable [8].

Finally there is the question of how *trans* fatty acids raise LDL and, in high concentration, also lower HDL. We [Abbey and Nestel, unpubl. data] have found that lipid transfer protein activity is increased; this leads to a transfer of cholesteryl esters from HDL to LDL, lowering the former and raising the latter.

References

1 Enig MG, Subodh A, Keeney M, et al: Isomeric *trans* fatty acids in the US diet. Am Coll Nutr 1990;9:471–486.
2 Hunter JE, Applewhite TH: Reassessment of *trans* fatty acid availability in the US diet. Am J Clin Nutr 1991;54:363–369.
3 Willett WC, Stampfer MJ, Manson J-A, et al: Intake of *trans* fatty acids and risk of coronary heart disease among women. Lancet 1993;341:581–585.
4 Mattson FH, Hollenback EJ, Kligman AM: Effect of hydrogenated fat on the plasma cholesterol and triglyceride levels of man. Am J Clin Nutr 1975;28:726–731.
5 Mensink RP, Katan MB: Effect of dietary *trans* fatty acids on high-density and low-density lipoprotein cholesterol levels in healthy subjects. N Engl J Med 1990;323:439–445.
6 Laine DC, Snodgrass CM, Dawson EA, et al. Lightly hydrogenated soy oil versus other vegetable oils as a lipid-lowering dietary constituent. Am J Clin Nutr 1982;35:683–690.
7 Nestel PJ, Noakes M, Belling GB, et al: Plasma cholesterol-lowering potential of edible-oil blends suitable for commercial use. Am J Clin Nutr 1992;55:46–50.
8 Nestel P, Noakes M, Belling B, et al: Plasma lipoprotein lipid and Lp(a) changes with substitution of elaidic acid for oleic acid in the diet. J Lipid Res 1992;33:1029–1036.
9 Troisi R, Willett W, Weiss ST: *Trans* fatty acid intake in relation to serum lipid concentrations in adult men. Am J Clin Nutr 1992;56:1019–1024.
10 Wood R, Kubena K, O'Brien B, et al: Effect of butter, mono- and polyunsaturated fatty acid-enriched butter, *trans* fatty acid margarine, and zero *trans* fatty acid margarine on serum lipids and lipoproteins in healthy men. J Lipid Res 1993;34:1–11.
11 Lichtenstein AH, Ausman LM, Carrasco W, et al: Hydrogenation impairs the hypolipidemic effect of corn oil in humans. Arterioscler Thromb 1993;13:154–161.

P.J. Nestel, MD, Chief, CSIRO Division of Human Nutrition, PO Box 10041,
Gouger Street, Adelaide, SA 5000 (Australia)

Galli C, Simopoulos AP, Tremoli E (eds): Fatty Acids and Lipids: Biological Aspects.
World Rev Nutr Diet. Basel, Karger, 1994, vol 75, pp 190–192

Alternatives for Nutritional *Trans* Fatty Acids

Ronald P. Mensink[1], *Gerard Hornstra*

Department of Human Biology, Limburg University, Maastricht, The Netherlands

Partial hydrogenation of vegetable oils is used to convert liquid oil into a semi-solid fat. This widely applied process leads to the formation of *trans* unsaturated fatty acids. The most abundant *trans* fatty acid in the human diet has 18 carbon atoms and 1 double bond (*trans*-C18:1). Controlled dietary studies have shown that, relative to *cis*-C18:1 (oleic acid), *trans*-C18:1 lowers the level of cholesterol in the anti-atherogenic high-density lipoproteins (HDL), but increases the level of cholesterol in the atherogenic low-density lipoproteins (LDL) and that of lipoprotein (a) (Lp(a)) [1–3]. A recent prospective epidemiological study found a positive relation between *trans*-C18:1 intake and the risk for coronary heart disease [4].

Sources and Intakes of *Trans*-C18:1

Hydrogenated oils are the major, but not the only source of *trans*-C18:1: fat from ruminant animals, such as cows and sheep, may also contain *trans*-C18:1. The contribution of *trans*-C18:1 from dairy products to total intake is about 20–40% [5, 6]. The actual intake, however, is difficult to estimate as most nutrient data bases do not contain values for *trans* fatty acids. However, using food availability data or duplicate portion analysis, 4–6% of total fatty acids do have the *trans* configuration, which corresponds with 1.5–3% of total energy intake. Intakes between individuals, however, may vary widely [7].

[1] R.P.M. is a research fellow of The Royal Netherlands Academy of Science (KNAW).

Alternatives for *Trans*-C18:1

Recently, we have carried out a double-blind cross-over study with 38 healthy men [8]. Subjects received for 6 weeks a Western-control diet; during a second period of 6 weeks, fat-containing items were substituted by a series of palm-oil-based products so that 70% of total fat intake was replaced. These products included margarines, frying and bakery fats, snack foods, cheese, ice cream, bakery products and chocolate spread. Diets were given in random order. The study was preceded by a run-in period of 3 weeks, and interrupted by a washout of 3 weeks. During these periods, subjects received the Western-control diet. In the fourth week of each experimental period, volunteers were instructed to collect duplicate portions of all foods and liquids consumed over a 48-hour period. Fasting blood samples were taken at 3-week intervals. The research protocol was approved by the Medical Ethical Committee of the University of Limburg, Maastricht. The Netherlands. Written informed consent was obtained from all participating volunteers.

According to duplicate portion analysis ($n = 35–38$) the intake (percent of total fatty acids; mean ± SEM) of myristic (4.7 ± 0.3 vs. 2.4 ± 0.1; $p < 0.001$), stearic (8.6 ± 0.2 vs. 6.9 ± 0.2; $p < 0.001$) and linoleic acid (12.7 ± 0.6 vs. 14.3 ± 0.6; $p < 0.001$) was lower, and that of palmitic acid (20.8 ± 0.6 vs. 28.6 ± 0.7; $p < 0.001$) higher on the palm-oil diet. The intake of oleic acid (37.2 ± 0.7 vs. 38.7 ± 0.4) was not different between the two dietary periods. The proportion of total fatty acids of *trans*-C18:1 was 4.7 ± 0.3% on the control diet and decreased to 2.1 ± 0.2% on the palm-oil diet ($p < 0.001$). The proportion of *trans*-C18:1 in the fatty acids of serum triglycerides decreased from 3.5 ± 0.1 to 2.8 ± 0.1% ($p < 0.001$), and that of platelet phospholipids from 1.0 ± 0.1 to 0.7 ± 0.1% ($p < 0.001$). After the first 6 weeks of the study, the level of *trans* in the diet correlated with the level of *trans*-C18:1 in serum triglycerides ($r = 0.41$; $p = 0.014$), and in platelet phospholipids ($r = 0.56$; $p < 0.001$). In addition, changes in dietary *trans*-C18:1 levels correlated with changes in the level of *trans*-C18:1 in serum triglycerides ($r = 0.56$; $p < 0.001$) and in platelet phospholipids ($r = 0.59$; $p < 0.001$). The palm-oil diet caused a highly significant decrease of 10% ($p = 0.006$) in the serum Lp(a) level, while the ratio between pro- and anti-atherogenic lipoproteins (LDL-HDL ratio) decreased by 8% from 2.90 ± 0.2 to 2.69 ± 0.2 ($p = 0.02$) [8, 9].

Conclusions

This study with healthy normocholesterolemic volunteers showed that analysis of the level of *trans*-C18:1 in serum triglycerides or platelet phospholipids is a sensitive marker to estimate the intake of *trans*-C18:1 in free-living populations. Also, our study showed that substitution of the habitual dietary fat by palm oil caused favorable changes in the serum lipoprotein profile. As the intake of more than one fatty acid changed, these effects were not necessarily due to a reduced intake of *trans*-C18:1. Therefore, specific studies should be designed to examine whether saturated fats with a relatively high stearic acid and/or palmitic acid level are a health-beneficial alternative for hydrogenated oils.

References

1 Mensink RP, Katan MB: Effect of dietary *trans* fatty acids on high-density and low-density lipoprotein cholesterol levels in healthy subjects. N Engle J Med 1990;323:439–445.
2 Mensink RP, Zock PL, Katan MB, et al: Effect of dietary *cis* and *trans* fatty acids on serum lipoprotein (a) levels in humans. J Lipid Res 1992;33:1493–1501.
3 Nestel P, Noakes M, Belling B, et al: Plasma lipoprotein lipid and Lp(a) changes with substitution of elaidic for oleic acid in the diet. J Lipid Res 1992;33:1029–1036.
4 Willett WC, Stampfer MJ, Manson JE, et al: Intake of *trans* fatty acids and risk of coronary heart disease among women. Lancet 1993;341:581–585.
5 Hunter JE, Applewhite TH: Reassessment of *trans* fatty acid availability in the US diet. Am J Clin Nutr 1991;54:363–369.
6 Gurr MI: *Trans* fatty acids – Metabolic and nutritional significance. Nutr Bull 1986;11:105–122.
7 Van den Reek MM, Craig-Schmidt MC, Weete JD, et al: Fat in the diets of adolescent girls with emphasis on isomeric fatty acids. Am J Clin Nutr 1986;43:530–537.
8 Sundram K, Hornstra G, van Houwelingen AC, et al: Replacement of dietary fat with palm oil: Effect on human serum lipids, lipoproteins and apolipoproteins. Br J Nutr 1992;68:677–692.
9 Hornstra G, van Houwelingen AC, Kester ADM, et al: A palm oil-enriched diet lowers serum lipoprotein (a) in normocholesterolemic volunteers. Atherosclerosis 1991;90:91–93.

Ronald P. Mensink, PhD, Department of Human Biology, Limburg University,
PO Box 616, NL-6200 MD Maastricht (The Netherlands)

Subject Index